Springer Theses

Recognizing Outstanding Ph.D. Research

Aims and Scope

The series "Springer Theses" brings together a selection of the very best Ph.D. theses from around the world and across the physical sciences. Nominated and endorsed by two recognized specialists, each published volume has been selected for its scientific excellence and the high impact of its contents for the pertinent field of research. For greater accessibility to non-specialists, the published versions include an extended introduction, as well as a foreword by the student's supervisor explaining the special relevance of the work for the field. As a whole, the series will provide a valuable resource both for newcomers to the research fields described, and for other scientists seeking detailed background information on special questions. Finally, it provides an accredited documentation of the valuable contributions made by today's younger generation of scientists.

Theses are accepted into the series by invited nomination only and must fulfill all of the following criteria

- They must be written in good English.
- The topic should fall within the confines of Chemistry, Physics, Earth Sciences, Engineering and related interdisciplinary fields such as Materials, Nanoscience, Chemical Engineering, Complex Systems and Biophysics.
- The work reported in the thesis must represent a significant scientific advance.
- If the thesis includes previously published material, permission to reproduce this must be gained from the respective copyright holder.
- They must have been examined and passed during the 12 months prior to nomination.
- Each thesis should include a foreword by the supervisor outlining the significance of its content.
- The theses should have a clearly defined structure including an introduction accessible to scientists not expert in that particular field.

More information about this series at http://www.springer.com/series/8790

Hazel Garvie-Cook

Novel (Trans)dermal Drug Delivery Strategies

Micro- and Nano-scale Assessments

Doctoral Thesis accepted by
the University of Bath, UK

Springer

Author
Dr. Hazel Garvie-Cook
Department of Physics
University of Bath
Claverton Down
UK

Supervisors
Prof. Sergey Gordeev
Department of Physics
University of Bath
Claverton Down
UK

Prof. Richard Guy
Department of Pharmacy & Pharmacology
University of Bath
Claverton Down
UK

ISSN 2190-5053 ISSN 2190-5061 (electronic)
Springer Theses
ISBN 978-3-319-28900-7 ISBN 978-3-319-28901-4 (eBook)
DOI 10.1007/978-3-319-28901-4

Library of Congress Control Number: 2015960828

Printed on acid-free paper

This Springer imprint is published by SpringerNature
The registered company is Springer International Publishing AG Switzerland

Parts of this thesis have been published in the following journal articles:

H. Garvie-Cook, K. Frederiksen, K. Petersson, R.H. Guy and S.N. Gordeev. Characterisation of topical film-forming systems using atomic force microscopy and Raman micro-spectroscopy. *Molecular Pharmaceutics*, 12(3):751–757, 2015.

H. Garvie-Cook, K. Frederiksen, K. Petersson, R.H. Guy and S.N. Gordeev. Biophysical elucidation of the mechanism of enhanced drug release and topical delivery from polymeric film-forming systems. *Journal of Controlled Release*, 212:103–112, 2015.

H. Garvie-Cook, J.M. Stone, F. Yu, R.H. Guy and S.N. Gordeev. Femtosecond pulsed laser ablation to enhance drug delivery across the skin. *Journal of Biophotonics*, 9(1–2):144–154, 2016.

AFM methodology and some images were included as an example in a patent filed by LEO Pharma A/S:

K. Petersson, K. Frederiksen, D. Omkvist, J. Jansson. A topical composition comprising a film-forming polymer for delivering an active ingredient to skin. WO2014006202 A1. 9th January 2014.

Supervisors' Foreword

Human skin is a remarkable example of sophisticated bioengineering. The skin has many vital functions including, most significantly, to limit the rate of water loss from the body and to protect the organism from exposure to harmful chemicals, viruses and bacteria. This impressive barrier function also constrains the diffusion of topically applied drugs into and across the skin. Nonetheless, there remains an obvious need to develop better and more efficient drug products to treat dermatological disease, and transdermal drug delivery—for the relief of local, subcutaneous inflammation and of various systemic conditions (e.g. analgesia, hormone replacement therapy, etc.)—presents, for the right compounds, very real advantages over more conventional oral and intravenous formulations.

Many methods have been developed to overcome skin's barrier function and to effectively deliver drugs at a sufficient rate to achieve the required therapeutic concentration at the site of action either within or below the skin, or systemically. Polymeric film-forming systems (FFS) represent a relatively novel class of formulations for (trans)dermal drug delivery. The FFS typically comprises a polymer, a plasticiser and a drug dissolved in a volatile solvent; upon application to the skin, a thin and flexible film is formed as the solvent evaporates. These films dry quickly, have good substantivity (i.e. resistance to being rubbed off), and are essentially invisible, features that surpass those of more conventional formulations, such as creams, ointments and gels, and confer good patient compliance. In addition, the area over which the film is applied can be varied, allowing flexible dosing to treat (for example) that part of the skin which is diseased.

In this thesis, Hazel Garvie-Cook investigated two FFS, one based on a relatively hydrophobic polymer, Eudragit, the other on the more hydrophilic Klucel. While atomic force microscopic (AFM) images of the films demonstrated the homogeneous incorporation of the plasticizer triethyl citrate, different behaviour was observed when the latter was replaced with medium-chain triglycerides (MCT), a commonly used excipient in dermatological formulations. In this case, AFM imaging and nanoindentation of the films revealed a two-phase structure with soft inclusions, 0.5 to 4 μm in diameter, surrounded by more rigid material. Chemical maps of the films, acquired using Raman micro-spectroscopy, revealed the spatial distribution of the constituents, and specifically that MCT was primarily confined within the inclusions. When a model drug, the corticosteroid, betamethasone valerate (BMV), was incorporated, its distribution was more uniform although, in contrast to MCT, somewhat more was found in the material surrounding the inclusions. The Raman spectra also showed that the physical state of the drug within the formed film depended on the polymer used, and upon whether it was located within the inclusions or within the surrounding material. Importantly, the biophysical characterisation of the FFS permitted the mechanism, by which the faster BMV release observed from formulations containing MCT (the more lipophilic plasticiser), to be explained.

Considerable research interest is currently focussed on methods to "porate" the skin, for example with microneedles, to facilitate the transdermal delivery of a much broader range of active compounds (including macromolecules, such as vaccines). Laser ablation of the skin has been considered for this purpose, but earlier efforts often produced thermal damage around the edges of the created pores, resulting in tissue coagulation and limited drug diffusion. More recently, however, it has been demonstrated that the thermal damage can be dramatically reduced by using very short (10^{-13} s) laser pulses that induce plasma-mediated ablation. In this case, the pulse duration is so short, and the skin so rapidly ablated, that no significant thermal damage to the surrounding tissue occurs. Unfortunately, such short laser pulses cannot propagate in standard optical fibres, making delivery of the laser radiation to the skin technically difficult. Hazel Garvie-Cook solved this problem by using hollow core fibres, designed and fabricated in the Department of Physics at the University of Bath, to specifically deliver ultrashort laser pulses to the skin. It was further demonstrated that the power threshold for poration could be dramatically reduced by the pre-application of ink to the skin and that, under these conditions, significantly increased skin permeation of a model drug was achieved. It follows that dye-enhanced, plasma-mediated ablation of the skin is a potentially new and advantageous approach with which transdermal drug delivery may be facilitated.

Claverton Down, UK
October 2015

Prof. Sergey Gordeev
Prof. Richard Guy

Abstract

The skin provides an effective barrier against the diffusion of drugs applied topically, to the surface of the skin. There are many advantages of this delivery route, however, which motivate its development. Formulation optimisation and methods to physically porate the skin are two methods discussed here to overcome this barrier and efficiently delivery drugs through the skin. Atomic force microscopy (AFM) and Raman micro-spectroscopy have been used complementarily to investigate topical polymeric films of varying compositions. AFM imaging showed the homogeneity of the incorporation of a commonly used plasticizer, triethyl citrate. A reduction in elastic modulus with increasing plasticizer content was determined using AFM nanoindentation. Raman chemical mapping of the plasticizer and drug, betamethasone-17-valerate (BMV), incorporated within the films revealed their even distribution, as suggested by AFM images. Medium-chain triglycerides (MCT) were incorporated as a putative plasticizer. AFM imaging and nanoindentation showed the formation of a two-phase system of softer inclusions surrounded by a more rigid film. More MCT, and less BMV, was contained within these inclusions than was found in their surroundings, as determined using Raman chemical mapping. The drug was equally soluble in all areas of Eudragit films with MCT, but was more soluble in the inclusions in Klucel films with MCT than their surroundings. The softer nature of the inclusions, and the environment they provide in which the drug is more soluble, were believed to contribute to the enhancement in release observed. A technique to porate the skin using femtosecond pulsed laser ablation has been developed. The application of dye to the surface of the skin reduced the power necessary for ablation and the resulting micropores showed little thermal damage, as determined using Raman micro-spectroscopy. Ablation was thought to be plasma mediated, with the ink providing a means to initiate the formation of the plasma. The delivery of caffeine across porated skin increased compared to intact skin, and the enhancement was greater when higher powers were used for poration.

Acknowledgements

First, many thanks to my supervisors Sergey Gordeev and Richard Guy for your constant support and encouragement. The interdisciplinary nature of this work has been really fascinating and I am very grateful for your engaged and interested supervision. Thank you very much for acting in my best interests and for being extremely helpful in your quick feedback in the last few weeks of my Ph.D.

Also, thanks to John Mitchels, Sarah Cordery, Wendy Lambson and Natalie Belsey for help with techniques along the way. Thanks to Jim Stone for your time and guidance in setting up a method for laser microporation.

Working with LEO Pharma A/S led to some interesting results and discussion and for that, and the opportunity to present at LEO, thanks to Kit Frederiksen and Karsten Petersson.

Finally, many thanks to Josh and to my wonderful family for your encouragement and constant support.

Acknowledgements

Contents

Abbreviations and Definitions

Acronyms

AFM	Atomic Force Microscopy
EBID	Electron Beam Induced Deposition
FFS	Film-Forming System
NA	Numerical Aperture
SEM	Scanning Electron Microscopy
TEC	Triethyl Citrate

Acronyms/Abbreviations and Definitions

Betamethasone-17-Valerate (BMV)	A drug commonly used for inflammatory skin conditions
Eudragit® RS PO (Eudragit)	A film-forming ammonio methacrylate copolymer type B
Klucel™LF (Klucel)	A film-forming hydroxypropyl cellulse
Medium-Chain Triglycerides (MCT)	Contain 6 to 12 carbon fatty acids
Photonic Crystal Fibre (PCF)	Optical fibre with the ability to guide femtosecond pulsed laser light
Root Mean Square Roughness (RMS Roughness)	Root mean square average of height deviations taken from the mean image data plan
Stratum Corneum (SC)	The uppermost layer of skin in mammals

Definitions

Bioavailability	Fraction of the administered dose which reaches the target tissue and the rate at which it gets there.
Chromophore	Components of tissue which absorb a specific wavelength of light. Examples in skin include collagen and water.
Dermal Delivery	Compounds are applied to the skin for delivery to pathological sites within the skin.
Excipient	Component of the vehicle, added to bulk up the formulation and stabilise the drug substance. An example of an excipient is glycerol.
Fluence	The energy per unit area per exposure to a laser beam.
Partition Coefficient	A measure of the relative affinity of a drug between two phases. In (trans)dermal delivery, the octanol and water partition coefficient is used as a guide for how well the drug will enter and cross the skin.
Percutaneous Absorption	Absorption through the skin from topical application.
Pulse duration	Also called pulse width or pulse length. The duration of time between the points that the intensity of the pulse is half its maximum.
Systemic	Affecting the entire body.
Topical Delivery	Compounds are applied to the surface of the skin and diffuse into/across the skin.
Transdermal Delivery	Compounds are applied to the skin for delivery to the blood stream and systemic circulation.
Vehicle	Formed from excipients and contains the drug substance. Together, the vehicle and the drug substance comprise the formulation.

Chapter 1
Introduction

The skin provides a vital barrier against the loss of water from the body and the entry of harmful foreign agents. The stratum corneum (SC), the uppermost layer of the skin, provides the majority of this barrier function. This thin layer (up to ~30 μm thick) also limits the penetration of drugs applied topically, to the surface of the skin. Although topical delivery presents practical difficulties, its advantages over other delivery routes, such as oral and intravenous delivery, motivate the ongoing development and optimisation of topical drug delivery strategies.

Formulations applied to the surface of the skin are designed for delivery to pathological sites within the skin (dermal delivery) or to the blood stream and systemic circulation (transdermal delivery). These formulations must be cosmetically acceptable, to ensure good patient compliance, and must also maximise the amount of drug available for absorption by the skin. Conventional topically applied formulations include creams, ointments and transdermal patches. These formulations are designed to provide a stable environment to contain the drug substance and to facilitate the required period and extent of its delivery into the skin.

Polymeric film forming systems (FFS) offer a novel and potentially advantageous approach to (trans)dermal delivery. A solution is applied to the skin and a thin and nearly invisible polymeric film forms upon solvent evaporation. The mechanical properties and release dynamics of these films can be altered using additives, such as plasticizers and lipid components. A good FFS would form a film which flexes with the skin when in situ on its surface and ideally the two would remain in intimate contact for a prolonged period of time without breaking.

The atomic force microscope (AFM) provides a means of nanoscale imaging and mechanical investigation of samples in physiological environments. Nanoindentation using AFM probe tips allows mechanical properties, such as elastic modulus, of nanoscale features in thin samples to be extracted. Atomic force microscopy has therefore proved especially useful in the investigation of biological and pharmaceutical samples, including drug particles, skin and skin creams. Samples are not stained or treated prior to investigation, allowing their behaviour during application to be predicted.

H. Garvie-Cook, *Novel (Trans)dermal Drug Delivery Strategies*,
Springer Theses, DOI 10.1007/978-3-319-28901-4_1

The development of Raman micro-spectroscopy has facilitated the chemical analysis of samples down to a resolution of approximately 1 μm. Based on the excitation of molecular vibrations, Raman spectroscopy of samples reveals fingerprint spectra of their constituents. The chemical distribution within samples can be mapped by acquiring a grid of spectra. The environment in which a substance is contained affects its Raman scattering, allowing its physical state to be determined. The state of a drug, be it crystalline or dissolved, for example, can be determined over a mapped area. Drug release from formulations is limited by the dissolution of crystalline material. To maximise the drug's bioavailability, ideally it would remain stable and dissolved within its formulation. Raman micro-spectroscopy can therefore be used to detect the formation of crystalline material within samples and hence has proved extremely useful in the assessment of pharmaceutical samples.

Regardless of the optimisation of a formulation, the penetration of drugs with molecular weights greater than 500 g mol^{-1}, such as vaccines and proteins, through the SC is severely limited. Several methods have therefore been developed to get around the barrier that the SC presents, permitting the diffusion of these larger compounds, whilst avoiding the entry of harmful pathogens. Micropores can be produced in the skin using laser ablation. Ideally, the skin around the ablated micropores would be free of thermal damage and would heal quickly after application of a drug substance. Devices are currently available which ablate the skin by rapidly heating water within the tissue using pulsed lasers. The vaporization of the water and the subsequent build up of pressure results in a micro-explosion, which removes tissue. The use of these devices is associated with adverse side effects, like skin irritation and prolonged healing time. The further development of this technique is therefore required to avoid these side effects and to provide an advantageous method of SC removal. Ablation using pico- and femtosecond pulsed lasers arises from a different mechanism to that used in the devices mentioned. Development in the manufacture of femtosecond pulsed lasers, and the ability to guide them using photonic crystal fibres (PCF), has facilitated investigation into their use for the ablation of skin. Their extremely short pulse duration, when compared to the thermal relaxation time of skin, and the mechanism by which they ablate the skin, present exciting advances in this field which could result in the development of more efficient and less damaging poration devices.

This thesis describes the complementary use of AFM and Raman micro-spectroscopy to assess and optimise polymeric films for (trans)dermal delivery of varying compositions. The addition of triethyl citrate (TEC), a commonly used plasticizer, and medium chain triglycerides (MCT), fatty acids with 6–12 carbon atoms, to films of two different polymers, Eudragit and Klucel, have been investigated. The homogeneity and elastic moduli of formed polymeric films were determined using AFM imaging and nanoindentation. Raman micro-spectroscopy provided complementary information on the chemical homogeneity of the additives and the drug substance within the films. The combination of the two techniques was used to elucidate the mechanism of drug release from the films. Femtosecond pulsed laser ablation of porcine skin was also investigated. The enhancement of delivery across the skin was determined using the in vitro release of caffeine. The application of ink to the surface of the skin prior to ablation reduced the laser power required and resulted in less thermal damage.

Background information and the current application of techniques relevant to this thesis are discussed in Chap. 2. This includes description of topical delivery and its optimisation, the principles behind AFM and Raman spectroscopy and their application in pharmaceutics, and the development of laser microporation. Details of the methodology used to acquire topographical, mechanical and chemical information on polymeric films are given in Chap. 3. The set-up and method of laser poration, as well as the technique used to determine the permeation of caffeine across the skin, are also described.

The incorporation of TEC plasticizer into polymeric films is discussed in Chap. 4. AFM images are presented, revealing the differing topographies of films of varying compositions. The deformation behaviour of the samples is shown in nanoindentation loops and the extraction of their elastic moduli using two different models, the elastic Hertzian model and an adapted Oyen & Cook model, are described. The results of Raman chemical mapping of films incorporating TEC are shown, revealing the chemical homogeneity of the formed films. Determination of the physical state of the incorporated drug within the films is also described.

Chapter 5 discusses the incorporation of MCT as a putative plasticizer. The use of AFM imaging to determine the nature of observed phase separation is described, as well as the nanoindentation of these heterogeneous films. Raman chemical maps show chemical disparity within these films and the physical state of the drug across the mapped areas. The mechanism behind the observed enhancement in release from films incorporating MCT is then discussed and attributed to the presence of the inclusions.

The poration of porcine skin with a femtosecond pulsed laser is described in Chap. 6. The refinement of the poration process is initially discussed and the use of a dye is described, to reduce the power required for ablation. Directing the laser beam towards the skin using a fibre resulted in the most reproducible results, which are given in this chapter. The Gaussian beam emitted by the laser is also modelled and used to explain the variation in pore dimensions with laser power and distance between the end of the fibre and the skin. The enhancement in caffeine permeation across porated skin is described and the mechanism of ablation discussed.

The results found in this work are summarised in Chap. 7 and conclusions are drawn. Further work is also suggested to develop and extend the techniques reported.

The work presented in this thesis describes the use of techniques to analyse novel (trans)dermal drug delivery strategies. AFM and Raman micro-spectroscopy investigations of polymeric films provide micro- and nano-scale information for the refinement of the composition of these formulations. Film-forming systems (FFS) can therefore be developed to remain in prolonged and intimate contact with the skin, making efficient use of the incorporated drug. Laser microporation of the skin has previously been achieved using lasers with nano- and microsecond pulse durations. Femtosecond pulsed lasers offer a novel alternative and their poration of the skin has not been extensively investigated. Ink applied to the surface of the skin reduces the power required for ablation, resulting in less thermal damage to the tissue surrounding the pores. The development of this laser system therefore presents a method of porating the skin in a short time with minimal thermal damage, enhancing the delivery of topically applied compounds and potentially minimising healing time.

Chapter 2
Background

2.1 Topical Drug Delivery

2.1.1 Structure of the Skin

The principal role of skin is to act as a barrier against evaporative water loss and against the entry of harmful foreign agents. The protective role of the skin is accomplished by its layered structure. The basic layers of human skin (outlined in Fig. 2.1a) consist of the hypodermis, epidermis and dermis. The hypodermis (up to several millimetres thick across the body) is made up of a network of fat cells and plays an important role in energy storage and metabolism [1]. It also provides protection against injury. The dermis, with thickness in the range of 0.3–3 mm, lies above the hypodermis and forms the bulk of the skin [2]. It is primarily composed of collagenous fibres and elastic connective tissue. This layer determines the elasticity of the skin and provides the physical support needed for networks of nerves and blood vessels [1, 3].

The epidermis (60–800 μm thick) contains distinct layers which are commonly categorised into the stratum corneum (SC), the uppermost layer, the stratum granulosum, the stratum spinosum and the stratum basale, the latter of which forms the lowest layer [3]. The viable epidermis, comprising of all layers apart from the SC, is primarily responsible for the formation of the SC and also contains melanocytes, which produce melanin for ultraviolet (UV) absorption, and Langerhans cells, which are responsible for the immune response of the skin [4]. After cell division in the stratum basale, cells migrate upwards towards the skin's surface. During this migration, the cells undergo differentiation, flattening and losing their nucleus. The SC is composed of approximately 10–15 layers of differentiated cells and this thin layer, up to approximately 30 μm thick, provides the majority of the skin's barrier function.

H. Garvie-Cook, *Novel (Trans)dermal Drug Delivery Strategies*,
Springer Theses, DOI 10.1007/978-3-319-28901-4_2

(a)

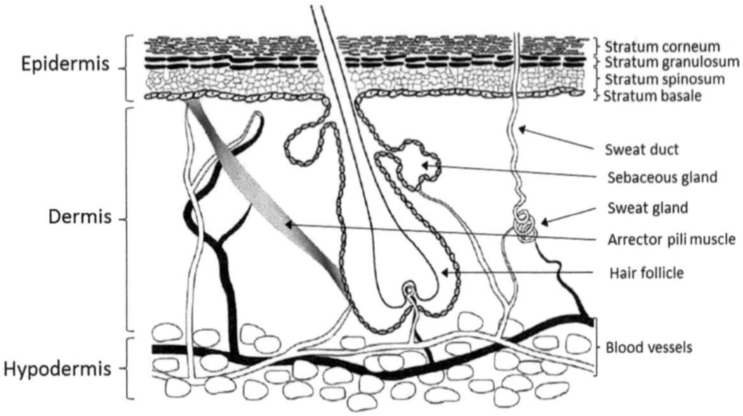

(b)

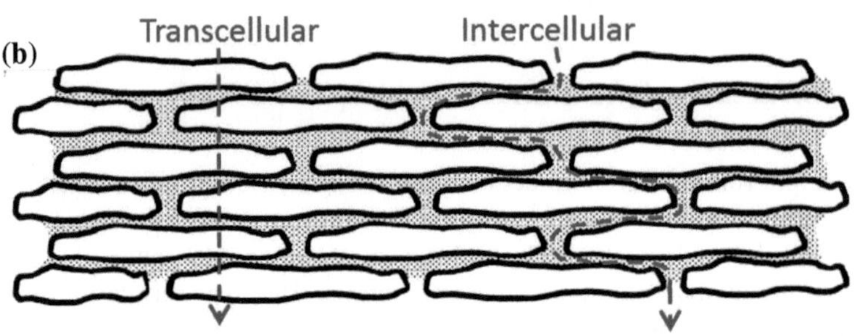

Fig. 2.1 **a** The structure of human skin. Image adapted from [3]. **b** The structure of the SC can be described using a "bricks and mortar" analogy. The corneocytes (in *white*) and the intercellular lipids (continuous domain in *dots*) represent the bricks and mortar, respectively. Two routes contributing to the percutaneous absorption of topically applied compounds, the transcellular and intercellular routes, are shown

2.1.1.1 Stratum Corneum

The SC has a "bricks and mortar" structure (Fig. 2.1b) of dead, anucleate cells, termed corneocytes, surrounded by a lipid matrix [5]. The SC comprises approximately 96 % proteins and water and 4 % lipids [6]. Corneocytes are primarily composed of bundled keratin surrounded by an envelope of cross-linked proteins and lipids. These flat, polygonal cells are typically 0.2–1.5 μm thick and have diameters of approximately 30–50 μm [3]. The composition of the intercellular lipids varies according to the individual and the anatomical site, but it mostly constitutes ceramides, fatty acids and cholesterol [7]. Corneodesmosomes are proteinaceous structures which connect corneocytes and contribute to the cohesion of the SC [1]. The hydration of the SC is maintained by the so-called natural moisturizing factor, a highly water soluble mixture of amino acids, derivatives thereof, urea and specific salts [8].

The barrier provided by the SC is influenced by physiological factors and skin disease. Age-related changes to the structure of the SC include the maturation of the underdeveloped premature neonatal skin [9] and an increase in corneocyte area and a decrease in SC thickness with increasing age [1]. Skin diseases, such as psoriasis and dermatitis, can affect both the protein and lipid composition of the SC, influencing and disrupting its barrier function [10]. UV exposure, hydration and chemical assault, amongst others, also affect the SC's barrier [3].

2.1.2 *Topical Medication*

Topical delivery refers to the application of formulations to the surface of the skin for the delivery of therapeutic agents (also known as drug substances) to pathological sites within the skin (dermal delivery) or through the skin into the blood stream and systemic circulation (transdermal delivery). Examples of dermatological conditions which are treated using dermal delivery are psoriasis and eczema [3], while the treatment of severe pain with fentanyl is an example of transdermal delivery [11]. In conventional topical therapy for dermatological disease, the drug is incorporated into a vehicle, such as a cream, which facilitates the application of the formulation and the delivery of the drug to the required site. The non-drug components of the formulation are termed excipients. The efficacy of topical delivery depends on the formulation and the diffusion of the therapeutic agent through the skin.

Transdermal and dermal routes of administration are advantageous in several aspects when compared, for example, to oral and intravenous delivery. For the oral treatment of skin disease, the systemic concentration of the drug must be high enough to achieve therapeutic benefit at the diseased site. In contrast, a lower dose is needed for a formulation that is applied directly to the affected site of the skin, decreasing or eliminating any adverse effects of the drug [12]. Transdermal delivery also avoids first pass metabolism of the drug, which decreases its concentration [13], allowing for a lower daily dose. Oral delivery involves a peak of drug concentration in blood and tissue, followed by a decline. (Trans)dermal delivery, however, can maintain drug level within the therapeutic window for a prolonged period of time, extending the duration of the action of the drug and reducing the frequency of dosing required. In the case of patches, the drug input can be terminated simply by removal of the patch.

2.1.2.1 Permeation Pathways

There are three main pathways by which drugs can permeate the SC: via the appendages (appendageal), through intercellular domains (intracellular) and through the cells themselves (transcellular) (Fig. 2.1b). The contributions of these three routes to the percutaneous absorption of a topically applied compound, from the surface of the skin either into the skin or into systemic circulation, depend on the nature of the permeating molecules and the density of appendages, such as hair follicles and sweat ducts, at the site of application [1].

Appendages penetrate the SC and the epidermis, providing a route for permeation which "bypasses" the barrier. The density of appendages varies, depending on the anatomical site, but is consistently low. This relatively low resistance route of transport plays a large role in iontophoris and is thought to be quite important for compounds of low SC diffusivity, such as hydrophilic and large molecular weight compounds [3].

The transcellular route requires the repeated penetration of the permeating compound into and out of the corneocytes and the intercellular lipids. In this route of permeation, the compound must therefore diffuse through both lipophilic regions (lipids) and hydrophilic regions (hydrated keratin within corneocytes) of the SC.

The major pathway for the permeation of small ($<500\,\mathrm{g\,mol^{-1}}$ [14]), uncharged molecules is believed to be intercellular [3]. Intercellular lipids, providing the only continuous phase within the SC, form lamellar structures between corneocytes. The permeation of compounds via this route is thought to occur by diffusion along and/or across lipid lamellae [1] which provide the rate limiting step [15].

2.1.2.2 Diffusion

Drug transport across the SC occurs by passive diffusion and can be described by Fick's first law [3],

$$J_{max} = \frac{DK_{SC/V}C_V{}^{sat}}{h} \tag{2.1}$$

where J_{max} is the maximum flux of the permeant, D is its diffusion coefficient in the barrier (typically the SC), $K_{SC/V}$ is the partition coefficient of the permeant between the SC and the vehicle, $C_V{}^{sat}$ is the saturation concentration of the permeant in the vehicle, and h is the diffusion pathlength. The diffusion coefficient measures how easily the permeant traverses the SC, while $K_{SC/V}$ describes the distribution of the permeant between the SC and the vehicle, and reflects the ratio of the compound's solubilities in these two phases,

$$K_{SC/V} = \frac{C_{SC}{}^{sat}}{C_V{}^{sat}} \tag{2.2}$$

where $C_{SC}{}^{sat}$ is the saturation concentration of the permeant in the SC. Under ideal circumstances, therefore, J_{max} is independent of the vehicle at its saturation concentration, assuming that the vehicle does not alter the properties of the SC or the permeant solubility in the SC. The diffusivity D of the drug through the SC can be altered by the diffusion of the vehicle's components into the SC, and the effect of this on the drug's solubility, and can be increased by incorporating an enhancer into formulations. The use of enhancers, however, may produce skin irritation. Drug flux can be maximised via the SC-vehicle partition coefficient and the concentration of the drug in the vehicle.

2.1.2.3 Formulation Considerations

The design and constituents of the formulation influence the onset, extent and duration of therapy. Topically applied formulations must both be cosmetically acceptable, to ensure patient compliance, and must also maximize the bioavailability of the drug in the formulation; that is, the fraction of the administered dose which reaches the target tissue and the rate at which it gets there. Topically applied formulations include creams, ointments, gels and lotions (all semi-solid), as well as transdermal patches (solid state). In the design of an effective formulation, the following must be taken into account: the physico-chemical properties of the incorporated drug, the stability of the excipients and the drug substance, and the cosmetic acceptability of the formulation (good feel on the skin, appropriate texture and fragrance) [3].

Examples of materials which are commonly used as dispersive media in the preparation of semi-solid formulations are oils and fats (e.g. petrolatum and triglycerides) and hydrogels (usually water and/or alcohol based). These ingredients are often included to stabilise, adhere, dilute or thicken the formulation [1]. Solvents form the basis of the formulation and a commonly used example is water, which is non-irritating and solubilises a wide range of compounds. Transdermal patches contain the drug within an adhesive layer or a (typically polymeric) matrix. The rate of delivery can be modified by varying the composition and structure of the patches.

Increasing the concentration of a compound within a vehicle increases the rate of its release (Eq. 2.1). The selection of a vehicle, however, is a compromise between the solubility of the compound and the SC-vehicle partition coefficient. If a vehicle is altered to make a compound more soluble, the SC-vehicle partition coeffiecient decreases (Eq. 2.2) and the "leaving tendency" of the drug from the vehicle is low. At saturation, the release of the compound from the formulation is generally maximum [1]. Below saturation, the rate of release of the compound is greatest for the vehicle in which it is least soluble (having the highest "leaving tendency"). Increasing the concentration of the drug above its saturation results in the formation of a suspension of the drug and smaller fractions of the compound are in solution. The period over which delivery at the maximum flux (provided by the solubilised drug within the over-saturated vehicle) is maintained becomes shorter, as the dissolution of crystalline material in the vehicle limits this process.

The stability of the drug within the formulation (the ability of the drug to maintain therapeutic properties during storage and use) is related to its solubility in the vehicle. Consideration of the metamorphosis of formulations post-application is vital: solvent loss through evaporation and/or uptake into the SC results in an increase of the concentration of the compound in the remaining formulation. This potentially leads to the formation of solid drug particles or crystals which are unlikely to be absorbed to the target site in the skin [16].

2.1.3 Polymeric Film-Forming Systems

Polymeric film-forming systems (FFS) are a potentially advantageous approach to topical delivery. Thin and nearly-invisible films are formed in situ on the skin upon application of the FFS and subsequent solvent evaporation. FFS comprise primarily a polymer, a drug substance and a volatile solvent, but may also include other excipients such as plasticizers and/or lipids [17, 18].

Polymeric FFS have previously been used, for example, as tissue glue for thread-free closing of incisions [19], as a preoperative skin preparation [20], and in ostomy care [21]. FFS for transdermal drug delivery of steroidal hormones [22, 23] and analgesics [24, 25] have been reported but there is limited literature describing their use in dermal delivery. Frederiksen et al. assessed the in vitro release of betamethasone-17-valerate (BMV), used in the treatment of inflammatory skin conditions, from polymeric films of varying compositions [18]. Release from the films was found to depend on the nature of the polymer and the incorporated plasticizer.

FFS present potential advantages over conventional dosage forms, such as higher dosing flexibility and better patient compliance [26]. While the area of a patch is fixed, that of a film can be determined by the patient, allowing flexible and complete coverage of the affected/target site. The nearly invisible appearance of the films renders them less noticeable and their rub off resistance and fast drying times are more acceptable than those of semi-solid preparations, such as creams [27].

2.1.3.1 Film Composition

Suitable polymers for the fabrication of film-forming systems must form clear, flexible films at temperatures close to that of the surface of the skin (approximately 28–32 °C). These film-forming polymers must also be soluble in a skin-tolerant, volatile solvent.

Increasing the polymer content of a film typically means that more drug can be incorporated. However, this means that the film becomes more viscous. While this is manageable up to a point, the viscosity must not be so high that the film's application from a spray is prevented [26].

Addition of plasticizer to the films increases their flexibility and their ability to "move" with the skin [26, 28]. More flexible films are less likely to crack, meaning that a constant area for drug transport is maintained over a prolonged period of time. Increasing plasticizer content also increases the adhesion of polymeric films [26], but can make them sticky on their outer surface. The type of plasticizer incorporated influences drug release from polymeric films [18, 29], as do other additives, such as lipids [17, 30] or penetration enhancers [26].

2.2 Atomic Force Microscopy

Atomic force microscopy (AFM) provides a 3D profile of the surface of a sample using forces between the sharp tip of an AFM and the sample surface. The probe tip is attached to the end of a sensitive cantilever. As the tip is scanned over the sample, the cantilever deflects according to the topography of the surface. A diagram of a basic AFM is shown in Fig. 2.2.

The main components of an AFM are a microscope stage (including the scanner, sample holder and a force sensor), control electronics and a computer. Typically, the cantilever deflection is measured using the reflected signal of a laser beam from the back of the cantilever. The laser is reflected to a four quadrant photodiode, which monitors the position of the beam by changes in its output voltage. Cantilevers are microfabricated from silicon or silicon nitride using photolithography techniques; probe tips typically have radii of curvature on the order of nanometers.

A piezoelectric device is used to adjust the vertical z position of the sample according to the feedback system being used. The sample is moved in the x, y plane using other piezoelectric devices. It is the z position of the sample, as a function of its x, y position, which forms the topographical AFM image.

There are several advantages of using the AFM over more conventional imaging approaches, such as light microscopy and scanning electron microscopy, especially for biological samples. Minimal treatment is required prior to imaging using the AFM, meaning that staining, labelling and coating of samples is avoided, minimizing damage to and alteration of the sample. Damage is also minimised in tapping

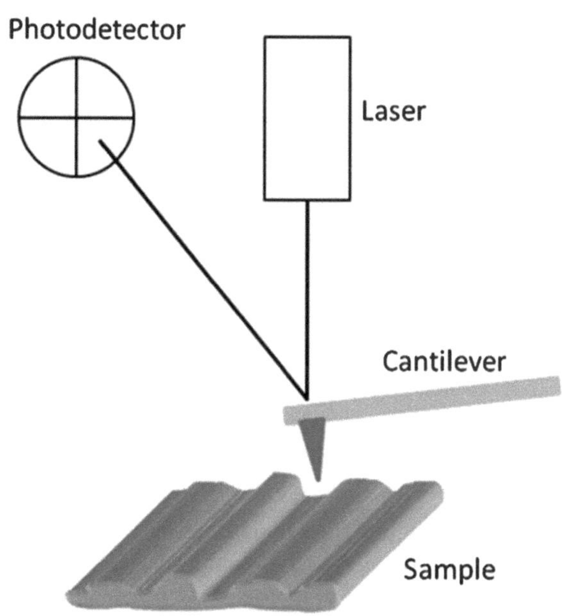

Fig. 2.2 Basic set-up of an AFM. The cantilever deflects according to interactions between the probe tip and the sample. This deflection is measured using the reflection of a laser beam from the back of the cantilever to a four quadrant photodiode

mode AFM imaging, further discussed in Sect. 2.2.1. The AFM produces 3D images, allowing for the entire surface structure of the sample to be better characterised. In addition, AFM can be used in physiologically relevant environments, facilitating live cell imaging and investigation of real time biological events at the nanoscale.

The major limitation of the AFM is its scan range, which is no more than approximately $100 \times 100\ \mu m^2$ in the x and y directions, and about $10\ \mu m$ in the z direction. The analysis of samples with roughness on or larger than this scale, such as skin, is therefore limited to rather small areas.

2.2.1 *Operating Modes for Imaging*

In contact mode AFM, the probe tip is always in contact with the sample. The cantilever is directly deflected by the surface of the sample and low stiffness cantilevers are used to maximise the deflection signal. A feedback system within the AFM corrects the height of the probe tip above the sample such that the cantilever deflection is maintained at a predetermined value, the set-point. The image of sample height is created from this change in cantilever height. Contact mode is most successful on stiff samples which do not deform under the load applied by the probe tip. However, the normal load from the tip and also the lateral force can distort and damage softer samples. Contact mode is also the most easily applied topographical mode when imaging samples in liquid [31].

There are several dynamic modes of operation where the cantilever is driven to oscillate, typically at its resonance frequency. As the oscillating probe approaches the sample, the interaction between the probe tip and the sample surface changes the frequency and amplitude of the oscillation. A feedback loop changes the height of the sample to maintain the amplitude or frequency of oscillation. It is from this change in height that a topographical image is acquired.

Intermittent-contact mode AFM is commonly used in ambient conditions. In this topographical mode, the cantilever is oscillated with a relatively large amplitude (1–100 nm), such that the tip and sample touch each other during each oscillation. The feedback system usually maintains the amplitude of the oscillation. As the cantilever is oscillated perpendicular to the surface of the sample, and the majority of the oscillation is out of contact with the sample, lateral forces are almost eliminated, making this an ideal topographical mode for imaging soft samples.

2.2.2 *Nanoindentation*

The AFM was designed initially as an imaging system, but the usefulness of the force sensitivity of the AFM cantilever has since been utilized in nanoindentation. In this technique, the response of the probe tip is recorded as it indents the sample. Material properties such as elastic modulus and hardness can be calculated at a spatial resolution limited by the size of the probe. Nanoindentation can also be carried out using dedicated apparatus but these measurements are usually on a larger scale with lower force sensitivity. AFM nanoindentation can be used to measure properties in thin and heterogeneous samples at the nanoscale. The interpretation of AFM nanoindentation data is further discussed in Sect. 3.1.3.

2.2.3 AFM Imaging and Nanoindentation in Pharmaceutics

The versatility of the AFM has facilitated the investigation of a large range of samples, such as metals, polymers, glasses and biological materials [31]. Little sample modification is required for AFM analysis and samples can be contained within physiologically relevant environments, meaning that this technique has proved extremely useful in the analysis of materials relevant to the pharmaceutical industry, including devices and drug particles.

Examples of the use of AFM for the analysis of pharmaceutical samples include investigation of the effects of humidity on spray-dried lactose that elucidated the process of crystallization [32]. The adhesion between lactose spheres and between compressed powder discs of lactose has also been used to predict the dispersion of these particles when delivered using dry powder inhalers [33]. Strong forces between particles have been shown to decrease the efficiency of these devices. The AFM has also been used to determine the size distribution of crystalline drug nanoparticles [34]. The resolution of the AFM permitted the detection of nanoparticles smaller than 40 nm in diameter (which are not detectable with the light scattering techniques commonly used). In these experiments, the AFM also provided information on the shape and structure of the nanoparticles that influence dosage performance. AFM imaging and nanoindentation have been used complementarily to distinguish nanoscale amorphous and crystalline domains on the surface of discs of crystalline sorbital (a commonly used excipient) [35]. The mechanism of drug release from solid lipid nanoparticles (SLN), during in vitro dissolution tests, was investigated using AFM imaging and nanoindentation [36]. Identification of soft, non-crystalline layers, surrounding the solid lipid core, was achieved by repeatedly imaging the particles. The particles appeared to spread out in this process, a phenomenum that was attributed to the deformation of the softer, outer layers by the AFM probe tip. Nanoindentation revealed the thickness of these outer layers and the initial, fast release of drug was explained by this softer, hydrophilic, non-crystalline outer layer.

Understanding the mechanics of the skin is useful for fully optimising topically applied drug delivery systems or devices for porating the skin. Solid-state dosage forms, such as the polymer films investigated in this work, ideally behave similarly to the skin when formed in situ on its surface. In this way, the two will flex together and remain in intimate contact for a prolonged period of time. The mechanical properties of the skin have previously been studied using cylindrical flat punches with radii ranging from 0.5–20 μm [37]. Elastic moduli extracted from these measurements decreased with increasing radius of the indenter. This was thought to be due to the softer, lower layers of skin bearing the load on the stiffer, uppermost layers. The elastic modulus extracted for the smallest radius probe (closest to the indenter geometry used in this work) was 0.03 (±0.02) GPa. The mechanical properties of corneocytes as a function of the depth from the SC surface were determined using nanoneedle AFM probes, with diameters between 30 and 80 nm [37]. This high resolution study of the mechanical structure of corneocytes revealed stiffer structures from within the cells and was predicted to be of value in the detection of disease-related changes in cells and the effects of, for example, UV exposure. The elastic modulus at indentations below 100 nm (surface elastic modulus) was approximately 0.3 GPa.

The effect of applying a commercially available moisturiser on the nanomechanics of skin has been determined [38]. The resistance of the skin to a scratch from an AFM probe tip was investigated and elastic moduli before and after the application of this skin cream were 0.09 (±0.03) and 0.05 (±0.02) GPa, respectively. The effect of moisturizer on individual corneocytes, isolated using tape stripping, has also been investigated and the surface roughness and elastic modulus of the cells shown to decrease [39].

2.3 Raman Spectroscopy

Raman spectroscopy yields information on molecular vibrations. These vibrations are sensitive to the strength and types of chemical bonds in the sample. Raman spectra can therefore be used to identify substances and help to elucidate their structures.

Raman spectroscopy is based on the scattering of incident light by molecules within a sample [40] (Fig. 2.3). Photons from the excitation light collide with molecules within the sample either elastically, without changing their energy (Rayleigh scattering), or inelastically, exchanging energy in the process (Raman scattering). When energy is exchanged, in Raman scattering, the molecule is excited to a virtual energy state. The molecule undergoes a transition from this virtual state to a lower energy state, which differs from its original state, and emits a photon.

If the frequency of the scattered photon is lower than that of the incident photon, the interaction is termed Stokes Raman scattering. Anti-Stokes scattering occurs

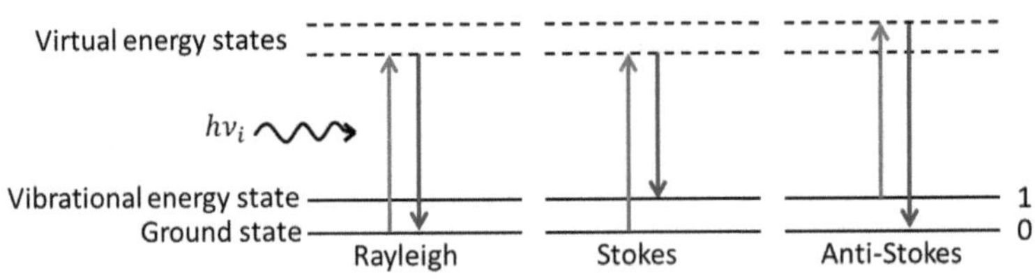

Fig. 2.3 Energy level diagrams showing Rayleigh and Raman (Stokes and anti-Stokes) scattering on excitation by a photon with energy $h\nu_i$

when the emitted photon has a higher frequency than the incident one. A molecule must be initially in an excited state for anti-Stokes Raman scattering to occur. Stokes Raman scattering arises from the interaction between a photon and a molecule in its ground state. Molecules are usually in their ground state at room temperature, so Stokes occurs more frequently than anti-Stokes scattering and is therefore more frequently used for spectroscopy.

The frequency shift of photons associated with their inelastic collisions is called spectral shift. A Raman spectrum shows the intensity of scattered radiation as a function of spectral shift and the intensity is directly proportional to the concentration of the scattering species [41].

Major components of a Raman spectrometer are the excitation source (a laser), the spectrometer, the detector and the optical set up [40]. The choice of laser depends on the sample being analysed. The use of particular wavelengths may excite strong, broad fluorescence in the samples, which can overwhelm the Raman signal. At higher excitation wavelengths, fewer fluorescent sources are excited but the intensity of Raman scattering is lower [42]. If samples are unlikely to fluoresce, shorter wavelengths in the visible range are usually employed, requiring a lower laser power and therefore reducing the likelihood of sample damage.

Raman spectroscopy is commonly compared to IR spectroscopy, which is based on the absorption, reflection and emission of light, as opposed to scattering. The two techniques can be used to provide complementary spectral information. In other instances, one technique may be preferred. For example, water has a weak Raman signal and does not interfere with the measurement of spectra from aqueous solutions [41]. The strong IR spectrum of water, however, may overwhelm the spectral information of the other components of an aqueous solution. The lateral spatial resolution of Raman spectroscopy is about 10-fold better than that of IR, (1 μm vs. 10 μm), facilitating the analysis of finer structure [43].

2.3.1 Raman Chemical Mapping

Chemical imaging combines spectroscopic techniques with optical microscopy. Spectra are acquired across the sample at predefined points. The total number of spectra in the map (corresponding to the total number of pixels) depends on the area of the map and the area over which individual spectra are acquired. High quality chemical maps are achieved using long exposure times and the acquisition of a large number of spectra. The collected spectra form a chemical image in which, for example, the concentration and the form of the constituents of the sample can be mapped. When analysing the data and forming a map, there is ideally at least one distinct peak for each of the constituents being mapped.

2.3.2 Raman Spectroscopy in Pharmaceutics

Since the development of instrumentation for Raman micro-spectroscopy, the method has been used, for example, to provide information on the homogeneity of the distribution of drug and excipients within a sample. This was demonstrated by mapping the distribution of estradiol within transdermal patches [44]. An inclusion of crystalline estradiol was observed (~250 μm in size), which was expected to negatively influence delivery and therefore compromise the efficacy of the patch. The distribution of drug and excipient within pharmaceutical tablets has also been shown [45], and related to the release rate. Raman spectroscopy has also been used to determine the physical state of a drug within its vehicle using the position, intensity and shape of characteristic spectral bands [46].

2.4 Laser Poration

The careful design of topically applied formulations can ensure the adequate delivery and therapeutic benefit of low molecular weight compounds. However, the penetration of compounds with molecular weights above 500 g mol^{-1}, for example vaccines and peptides, is severely limited by the SC barrier, regardless of formulation optimisation [14]. This is unfortunate in the case of vaccines because their delivery into viable skin is expected to elicit a more efficient immune response than that seen when they are injected subcutaneously or intramuscularly [47, 48]. It has therefore been argued that this deficiency, as well as mitigating the risks associated with injections, such as needle re-use and pain, might be avoided by delivery through porated skin, the SC of which has been superficially ablated with a laser, for example. Such an approach would also enhance the delivery of small molecular weight compounds [49].

2.4.1 SC Penetration Enhancement Methods

The SC barrier may be circumvented, at least partially, by formulation optimisation, energy driven approaches, minimally invasive technology and SC removal [50]. The latter can be achieved using radio-frequency ablation, laser microporation and thermal ablation, for instance [51].

Microporation using laser irradiance to ablate the SC provides several potential advantages. It is a needle-free technique, which does not produce bio-hazardous waste, and has been reported to be relatively painless [52]. The same device for poration can be used repeatedly, in contrast to the use of microneedle patches, for example. The area and depth of ablated pores depend on laser power and its delivery to the skin, ensuring that the approach is controllable and predictable [47, 53].

Furthermore, by limiting thermal damage to a small area, the pores produced heal quickly [47]. Devices are currently available to produce micropores in the skin and enhance the delivery of topically applied drugs [54, 55].

2.4.2 Laser Ablation

Tissue ablation is any process of tissue incision or removal. Laser ablation of skin is the removal of tissue by laser irradiance. The mechanism by which the tissue is removed depends on the wavelength and pulse duration of the laser used and on tissue characteristics [56]. There are three major models of tissue ablation by short laser pulses: selective photothermolysis and its variations (photothermal and photomechanical); photochemical; and plasma-mediated [57]. Earlier investigations of tissue ablation used lasers with nanosecond or greater pulse lengths, for which photothermal and photomechanical mechanisms (when visible or infrared wavelengths were used [58, 59]) or a photochemical mechanism (in the case of UV ablation [60]) were believed to occur.

Photothermal ablation is achieved by rapid heating of a target chromophore within the tissue [61]. Chromophores are components of tissue which absorb a specific wavelength of light and examples include collagen, melanin and water [56]. The target chromophore absorbs laser radiation and heats the water in the surrounding tissue. The subsequent vaporization of the water leads to a build-up of pressure within the tissue and a micro-explosion occurs, which ablates material from the surroundings [48]. In the case of short laser pulses, their duration can be less than the thermal relaxation time of the targeted tissue component, resulting in little heat transfer, and therefore less damage, to the surrounding tissue [62]. Tissue can also be ablated photomechanically, when the stress induced by laser heating exceeds the strength of the material. This leads to ejection of tissue fragments [63].

In photochemical ablation, the removal of tissue also results from explosions caused by the build-up of pressure. The pressure in this case derives from the photochemical dissociation of macromolecular bonds [60]. For far-UV (<200 nm) laser irradiation, the photon energy is larger than that of the chemical bonds in the molecule [64] and absorption of photons provokes electronic excitation and subsequent decomposition of the tissue into smaller, volatile fragments.

For pulse lengths in the range of picoseconds to femtoseconds, plasma-mediated ablation has been proposed [56, 65, 66]. Laser induced breakdown is one mechanism for creating plasmas. In this process, the tissue under irradiance is partially or completely ionized through absorption of laser energy and this results in a "gas" of charged particles, a plasma [67]. Plasmas absorb optical radiation much more strongly than ordinary matter and therefore limit the penetration of the incident light [68]. Once the pulse has terminated, free electrons recombine with the positively ionized molecules within the plasma and energy is transferred to surrounding material which is ablated. The energy from this recombination can also be converted to

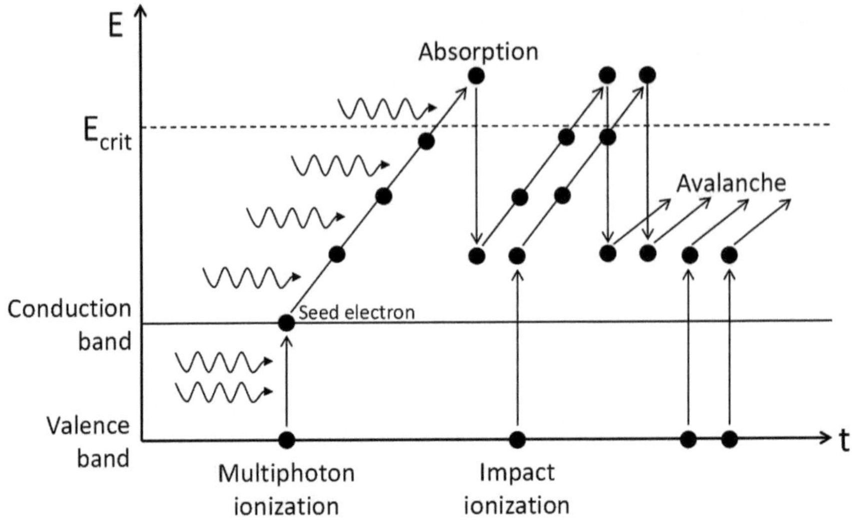

Fig. 2.4 Multiphoton ionization, absorption and impact ionization leading to an avalanche growth in the number of free electrons and the formation of plasma. E_{crit} is the critical electron energy for impact ionization to occur. Adapted from [66]

mechanical energy, resulting in the propagation of a shock wave and the formation of a cavitation bubble [69].

An initial threshold electron density must be reached for the formation of the plasma, and the threshold for laser induced breakdown is an electron density of 10^{18}–10^{20} cm^{-3} in the focal volume [65]. Two mechanisms can produce electron densities in this range (Fig. 2.4): (i) avalanche (or cascade) ionization, initiated either by seed electrons from impurities or by multiphoton absorption, and (ii) pure multiphoton ionization.

Free electrons are required in the focal volume for avalanche ionization to occur [65]. These "seed" electron(s) absorb photons, accelerate and then release bound electrons by collisional ionization. The freed electrons are then accelerated, freeing more electrons and the avalanche process continues, resulting in the formation of plasma. The "seed" electrons required for this process may come from easily ionized impurities or from multiphoton ionization.

Multiphoton ionization occurs when electrons are freed by absorption of multiple photons. High electron densities are produced by pure multiphoton ionization when pulse lengths are in the low femtosecond range [65] so only instantaneous ionization can occur in the duration of the time that the pulse is present in the focal volume. Each atom is independently ionized so no seed electrons, collisions or particle-particle interactions are required.

2.4.3 Laser Poration of Skin

The potential of lasers in biomedical applications was realised soon after the invention of the pulsed ruby laser in 1960 [70]. It was predicted that lasers could provide previously unachievable precision and selectivity in the manipulation of tissues [71]. Within a couple of years, the effect of exposure of skin to a pulsed ruby laser beam had been investigated. Goldman et al. [72] exposed non-pigmented and pigmented rabbit and human skin to a pulsed ruby laser beam ($\lambda = 694.3$ nm) with a pulse length of 200 μs and observed an enhanced interaction between the laser and the pigmented skin. Subsequent experiments showed that tissue changes were related to the skin pigmentation, the type of laser used, the fluence (the energy per unit area per exposure) and the duration of the exposure to the laser beam [73, 74]. The possible treatment of birthmarks (nevi), melanomas and tattoos, based on these findings, was proposed [75].

The first demonstration of controlled removal of the SC to enhance percutaneous transport was in 1987 [76]. In this study, an excimer laser was used ($\lambda = 193$ nm) with a pulse length of 14 ns. An unexpected decrease in permeation when using a higher laser fluence to remove the SC was observed, however, even though similar amounts of material were removed. It was concluded that tissue damage and cauterization occurred when a higher fluence was used, due to the build-up of heat. Ideally, for drug delivery applications, the pore created would not be surrounded by coagulated tissue, as this effectively blocks subsequent permeation through the pores [54].

Shortly thereafter, it was demonstrated that nanosecond pulsed lasers, with wavelengths strongly absorbed by the water in the skin, caused much less thermal damage of the surrounding tissue [77, 78]. Since then, a range of laser systems have been used for SC removal, as illustrated in Table 2.1.

Over the last 25 years, laser ablation of skin has focussed on the wavelengths of 10,600 and 2940 nm, produced using CO_2 and Er:YAG lasers, respectively. This causes ablation of the skin photothermally using the pulse durations reported in Table 2.1. Here, conventional ablation refers to the use of just one, large (mm in size) pore for drug permeation enhancement or cosmetic benefit (such as skin resurfacing or wrinkle treatment [93]). The use of these lasers, however, was associated with delayed re-epithelialisation and erythema (skin redness), with long recovery times of the large wounds produced [94].

Fractional photothermolysis, first described in 2004 [95], involves the use of a split laser beam to create an array of micropores in the skin. Fractional lasers are advantageous in that they create smaller zones of damage, within the surrounding, healthy skin, that are quicker to heal [82, 87].

The hand-held P.L.E.A.S.E.® device (Precise Laser Epidermal System), developed by Pantec Biolsolutions AG (Ruggell, Liechtenstein), uses a fractional Er:YAG laser to create micropores in the skin for both drug delivery enhancement [47, 53, 83–86] and cosmetic procedures [96]. The number of pores and the fluence used to create the pores are specified prior to the use of the device, enabling the distribution

Table 2.1 Illustrative articles reporting laser ablation of skin to enhance the permeation of topically applied substances

Year	Laser (wavelength, pulse duration)	in vitro/in vivo	Drug/permeant
1987	Conventional excimer (193 nm, 14 ns)	in vitro, human	Tritiated water [76]
1991	Conventional Er:YAG (2970 nm, 250 µs)	in vitro, porcine	Hydrocortisone[a], interferon[b] [58]
2002	Conventional ruby (694 nm, 40 ns), Er:YAG (2940 nm, 250 µs), CO_2 (10,600 nm, 50 ms)	in vivo, murine	5-Fluorouracil[c] [79]
2006	Conventional Er:YAG (2940 nm, 600 µs)	in vivo, human	Lidocaine[d] [80]
2008	Conventional Nd:YAG (1064, 532 and 355 nm, pulse durations of 11, 11 and 18 ns, respectively)	in vitro, rabbit	5-Fluorouracil[c] [81]
2010	Fractional and conventional Er:YAG (2940 nm, 350 µs)	in vitro, porcine	ALA[e] [82]
2010–2013	Fractional Er:YAG (P.L.E.A.S.E.®, 2940 nm, pulse duration unknown but photothermal ablation reported)	in vitro/in vivo, porcine, murine and human	Lidocaine[d] [83], prednisone[f] [53], ovalbumin[g] [47], diclofenac[h] [84], antibodies (ATG and basiliximab)[i] [85], Proteins (12.4–70 kDa) [86]
2010	Fractional Er:YAG (2940 nm, 400 µs)	in vitro/in vivo, murine/porcine	ALA[e] [87]
2010	Fractional CO_2 (10,600 nm, 3 ms)	in vitro/in vivo, porcine	Methyl ALA[e] [55]

(continued)

Table 2.1 (continued)

Year	Laser (wavelength, pulse duration)	in vitro/in vivo	Drug/permeant
2011	Conventional frequency doubled Nd:YAG (532 nm, 11 ns)	in vitro, rabbit	ALA[e] [59]
2012	Fractional CO_2 (10,600 nm, 6 ms)	in vitro, human	Polyethylene glycols (240–4300 Da)[j] [88]
2012	Fractional CO_2 (Ultrapulse, 10,600 nm, 5 ms)	in vivo, murine	Ovalbumin[g] [89]
2013	Fractional CO_2 (10,600 m, 16 ms)	in vivo, human	ALA[e] [90]
2013	Ti:Sapphire (801 nm, 70 fs)	in vivo, murine	ALA[e] [91]

[a]Hydrocortisone is a topical steroid for the treatment of inflammatory skin conditions
[b]Interferons are proteins which boost the immune system [92]
[c]5-Fluorouracil is an anticancer agent [79]
[d]Lidocaine is a common local anaesthetic [84]
[e]5-Aminolevulinic acid (ALA) and methyl ALA are photosensitizers used in photodynamic therapy [55, 59]
[f]Prednisone is a steroid used, for example, as an immunosuppressant [53]
[g]Ovalbumin is a model vaccine antigen [89]
[h]Diclofenac is a nonsteroidal anti-inflammatory drug [84]
[i]Anti-thymocyte globulin (ATG) and Basiliximab are proteins used for immunosuppression [85]
[j] Polyethylene glycols (PEGs) were used as high molecular weight model compounds [88]

and depth to be tailored for the application [47]. The pores created using this device are approximately 150–200 μm in diameter [47].

Adverse effects after ablation by fractional Er:YAG and CO_2 lasers include skin irritation and swelling, as well as those discussed for conventional laser ablation [55, 87, 97, 98]. Scarring has also been associated with the use of CO_2 lasers. Fewer of these symptoms are reported when fractional lasers are used [99]. The healing time after fractional or conventional laser treatment is also a concern, especially when considering multiple long-term treatments. Cost and the complex design of poration systems are considered drawbacks to their use for the enhancement of percutaneous transport [54].

The adverse side effects observed in ablation with microsecond pulses could be addressed with the use of ultra-short pulses, in the pico- or femto-second range. When these short pulses are used for tissue ablation, the laser energy is deposited and quickly transferred to kinetic energy of the ejected material, with little being transferred to the surrounding tissue, and minimal, if any, thermal damage [100]. Ideally, the laser system would be cheaper to develop and operate. The development of the technology associated with a device for ablation (cheaper laser, fibre optics) would bring down the cost of this kind of system. Also, the safety of a device, which uses a laser, can be improved with the use of visible light. As the latter can be detected by eye, it is much easier to avoid any unintentional exposures and harmful effects.

2.4.3.1 Ablation with Visible Wavelengths and Femtosecond Pulses

There have been few reported uses of visible laser light (400–700 nm) poration for enhanced drug delivery. At visible wavelengths, melanin and haemoglobin are the skin chromophores that have the greatest absorption capabilities. This means that the absorption of visible laser light at longer pulse lengths by skin depends on its pigmentation. Visible wavelengths are used, for example, in tattoo removal [101] and the treatment of pigmented lesions [62]. The use of a frequency doubled Nd:YAG laser (λ = 532 nm) with nanosecond pulses has been reported to enhance the delivery of 5-fluorouracil [81] and 5-aminolevulinic acid (ALA) [59]. In the first of these studies [81], irradiation at 532 nm was found to enhance the delivery of 5-fluorouracil, compared to its penetration across untreated skin, with no visible damage to the surrounding skin. At a given fluence, the enhancement at wavelengths of 355 and 1064 nm was greater, but damaging effects on the tissue surrounding the pores were observed. In a separate study, a lower fluence from an Er:YAG laser was required to enhance ALA delivery compared to that needed by a frequency doubled Nd:YAG laser with a wavelength of 532 nm [59]. Histological studies again showed no damage to the skin was caused by irradiation with the 532 nm laser.

Pulse lengths on the order of tens of femtoseconds have been used to ablate, for example, the cornea [102], teeth [103], skin [104, 105] and neural tissue [106]. The plasma mediated ablation, which occurs when ultrashort (pico- and femto-second) pulse durations are used, is often associated with the generation of stress that causes tissue damage [107]. In principle, the expansion of plasma, formed by laser irradiance, leads to the propagation of a shock wave through the tissue. However, the shock wave produced by femtosecond pulses has extremely high frequency and cannot propagate through tissue, being entirely absorbed in a subsurface layer less than 1 μm in thickness [107]. There is a paucity of literature describing the use of femtosecond pulsed lasers to enhance the permeation of therapeutic agents into and across the skin. Nicolodelli et al. [91] porated skin using a Ti:Sapphire laser (λ = 801 nm) with a pulse length of 70 fs and found that this enhanced the depth of ALA in the skin. Thermal damage, however, was observed surrounding the pores.

The optical absorption of the skin at 532 nm is low [56]. Greater fluences are therefore required for laser induced breakdown as seed electrons are not so easily produced at this wavelength. The absorption of the green light can be enhanced by applying a strongly absorbing dye to the surface of the skin. This effect has been investigated [108] using pulse lengths from 50 to 200 ms. Carbon particles mixed with film forming polymers provided the means by which green light was absorbed and converted to thermal energy. The presence of the dye induced tissue ablation but thermal injury was observed in a band of collagen denaturation in the surrounding tissue. The use of carbon suspensions to enhance laser treatment of acne and facial pores with a 1064 nm laser [109, 110] has been reported but there are no examples of its use to enhance drug delivery.

There has been no reported use of ablation of dyed skin using a visible laser with femtosecond pulses. The mechanism of ablation deriving from this combination is of interest to the efficient delivery of vaccines, proteins, peptides and smaller drug molecules into and across the skin.

References

1. H. Schaefer, T.E. Redelmeier, *Skin Barrier: Principles of Percutaneous Absorption* (Karger, Basel, New York, 1996)
2. D.H. Kim, N.S. Lu, R. Ma, Y.S. Kim, R.H. Kim, S.D. Wang, J. Wu, S.M. Won, H. Tao, A. Islam, K.J. Yu, T.I. Kim, R. Chowdhury, M. Ying, L.Z. Xu, M. Li, H.J. Chung, H. Keum, M. McCormick, P. Liu, Y.W. Zhang, F.G. Omenetto, Y.G. Huang, T. Coleman, J.A. Rogers, Epidermal electronics. Science **333**(6044), 838–843 (2011)
3. A. Williams, *Transdermal and Topical Drug Delivery from Theory to Clinical Practice* (Pharmaceutical Press, London, 2003)
4. C. Ehrhardt, K. Kim, Drug Absorption Studies : in Situ, in Vitro and in Silico Models. Biotechnology : Pharmaceutical Aspects (Springer, New York, 2008)
5. P.M. Elias, Epidermal lipids, membranes, and keratinization. Int. J. Dermatol. **20**(1), 1–19 (1981)
6. R.J. Scheuplein, A personal view of skin permeation (1960–2013). Skin Pharmacol. Physiol. **26**(4–6), 199–212 (2013)
7. R.H. Guy, Skin—'that unfakeable young surface'. Skin Pharmacol. Physiol. **26**(4–6), 181–189 (2013)
8. A.V. Rawlings, I.R. Scott, C.R. Harding, P.A. Bowser, Stratum-corneum moisturization at the molecular-level. J. Invest. Dermatol. **103**(5), 731–740 (1994)
9. N. Sekkat, Y.N. Kalia, R.H. Guy, Porcine ear skin as a model for the assessment of transdermal drug delivery to premature neonates. Pharm. Res. **21**(8), 1390–1397 (2004)

10. K.C. Madison, Barrier function of the skin: "la raison d'etre" of the epidermis. J. Invest. Dermatol. **121**(2), 231–241 (2003)
11. M.B. Delgado-Charro, R.H. Guy, Effective use of transdermal drug delivery in children. Adv. Drug Delivery Rev. **73**, 63–82 (2014)
12. V.V. Ranade, J.B. Cannon, *Drug Delivery Systems*, 3rd edn. (CRC Press, Boca Raton, 2011)
13. M.R. Prausnitz, R. Langer, Transdermal drug delivery. Nat. Biotech. **26**(11), 1261–1268 (2008)
14. J.D. Bos, M.M.H.M. Meinardi, The 500 dalton rule for the skin penetration of chemical compounds and drugs. Exp. Dermatol. **9**(3), 165–169 (2000)
15. R. Scheuplein, L. Ross, Effects of surfactants and solvents on the permeability of epidermis. J. Soc. Cosmet. Chem. **21**(13), 853–873 (1970)
16. B.G. Saar, L.R. Contreras-Rojas, X.S. Xie, R.H. Guy, Imaging drug delivery to skin with stimulated raman scattering microscopy. Mol. Pharm. **8**(3), 969–975 (2011)
17. N.B. Shelke, M. Sairam, S.B. Halligudi, T.M. Aminabhavi, Development of transdermal drug-delivery films with castor-oil-based polyurethanes. J. Appl. Polym. Sci. **103**(2), 779–788 (2007)
18. K. Frederiksen, R.H. Guy, K. Petersson, Formulation considerations in the design of topical, polymeric film-forming systems for sustained drug delivery to the skin. Eur. J. Pharm. Biopharm. **91**, 9–15 (2015)
19. M. Donkerwolcke, F. Burny, D. Muster, Tissues and bone adhesives—historical aspects. Biomaterials **19**(16), 1461–1466 (1998)
20. D.K. Jeng, A new, water-resistant, film-forming, 30-second, one-step application iodophor preoperative skin preparation. Am. J. Infect. Control **29**(6), 370–376 (2001)
21. J. Stephen-Haynes, C. Stephens, Evaluation of clinical and financial outcomes of a new no-sting barrier film and barrier cream in a large uk primary care organisation. Int. Wound J. **10**(6), 689–696 (2013)
22. I.Z. Schroeder, P. Franke, U.F. Schaefer, C.M. Lehr, Delivery of ethinylestradiol from film forming polymeric solutions across human epidermis in vitro and in vivo in pigs. J. Controlled Release **118**(2), 196–203 (2007a)
23. A. Misra, R.S. Raghuvanshi, S. Ganga, M. Diwan, G.P. Talwar, O. Singh, Formulation of a transdermal system for biphasic delivery of testosterone. J. Controlled Release **39**(1), 1–7 (1996)
24. C. Padula, G. Colombo, S. Nicoli, P.L. Catellani, G. Massimo, P. Santi, Bioadhesive film for the transdermal delivery of lidocaine: in vitro and in vivo behavior. J. Controlled Release **88**(2), 277–285 (2003)
25. H.O. Ammar, M. Ghorab, A.A. Mahmoud, T.S. Makram, A.M. Ghoneim, Rapid pain relief using transdermal film forming polymeric solution of ketorolac. Pharm. Dev. Technol. **18**(5), 1005–1016 (2013)
26. I.Z. Schroeder, P. Franke, U.F. Schaefer, C.M. Lehr, Development and characterization of film forming polymeric solutions for skin drug delivery. Eur. J. Pharm. Biopharm. **65**(1), 111–121 (2007b)
27. X. Tan, S.R. Feldman, J.W. Chang, R. Balkrishnan, Topical drug delivery systems in dermatology: a review of patient adherence issues. Expert Opin. Drug Delivery **9**(10), 1263–1271 (2012)
28. J.W. McGinity, L.A. Felton, *Aqueous Polymeric Coatings For Pharmaceutical Dosage Forms Drugs and the pharmaceutical sciences*, 3rd edn. (Informa Healthcare, New York, 2008)
29. F. Lecomte, J. Siepmann, M. Walther, R.J. MacRae, R. Bodmeier, Polymer blends used for the aqueous coating of solid dosage forms: importance of the type of plasticizer. J. Controlled Release **99**(1), 1–13 (2004)
30. D. Lunter, R. Daniels, In vitro skin permeation and penetration of nonivamide from novel film-forming emulsions. Skin Pharmacol. Physiol. **26**(3), 139–146 (2013)
31. P.J. Eaton, P. West, *Atomic Force Microscopy* (Oxford University Press, Oxford, New York, 2010)

32. R. Price, P.M. Young, Visualization of the crystallization of lactose from the amorphous state. J. Pharm. Sci. **93**(1), 155–164 (2004)
33. M.D. Louey, P. Mulvaney, P.J. Stewart, Characterisation of adhesional properties of lactose carriers using atomic force microscopy. J. Pharm. Biomed. Anal. **25**(3–4), 559–567 (2001)
34. H.Q.G. Shi, L. Farber, J.N. Michaels, A. Dickey, K.C. Thompson, S.D. Shelukar, P.N. Hurter, S.D. Reynolds, M.J. Kaufman, Characterization of crystalline drug nanoparticles using atomic force microscopy and complementary techniques. Pharm. Res. **20**(3), 479–484 (2003)
35. S. Ward, M. Perkins, J.X. Zhang, C.J. Roberts, C.E. Madden, S.Y. Luk, N. Patel, S.J. Ebbens, Identifying and mapping surface amorphous domains. Pharm. Res. **22**(7), 1195–1202 (2005)
36. Z. zur Muhlen, E. zur Muhlen, H. Niehus, W. Mehnert. Atomic force microscopy studies of solid lipid nanoparticles. Pharm. Res., **13**(9), 1411–1416 (1996)
37. M.L. Crichton, X.F. Chen, H. Huang, M.A.F. Kendall, Elastic modulus and viscoelastic properties of full thickness skin characterised at micro scales. Biomaterials **34**(8), 2087–2097 (2013)
38. B. Bhushan, W. Tang, S. Ge, Nanomechanical characterization of skin and skin cream. J. Microsc. **240**(2), 135–144 (2010)
39. R.M. Gaikwad, S.I. Vasilyev, S. Datta, I. Sokolov, Atomic force microscopy characterization of corneocytes: effect of moisturizer on their topology, rigidity, and friction. Skin Res. Technol. **16**(3), 275–282 (2010)
40. S. Sasic, *Pharmaceutical Applications of Raman Spectroscopy. Wiley Series on Technologies for the Pharmaceutical Industry*. (Wiley-Interscience, Hoboken, 2008)
41. A.C. Williams, H.G.M. Edwards, B.W. Barry, Fourier-transform raman-spectroscopy—a novel application for examining human stratum-corneum. Int. J. Pharm. **81**(2–3), R11–R14 (1992)
42. C.M. McGoverin, T. Rades, K.C. Gordon, Recent pharmaceutical applications of raman and terahertz spectroscopies. J. Pharm. Sci. **97**(11), 4598–4621 (2008)
43. F.C. Clarke, J.M. Jamieson, D.A. Clark, S.V. Hammond, R.D. Jee, A.C. Moffat, Chemical image fusion. the synergy of ft-nir and raman mapping microscopy to enable a more complete visualization of pharmaceutical formulations (vol 73, p 2157, 2001). Anal. Chem. **73**(10), 2369–2369 (2001)
44. G.L. Armstrong, H.G.M. Edwards, D.W. Farwell, A.C. Williams, Fourier transform raman microscopic study of drug distribution in a transdermal drug delivery device. Vib. Spectrosc. **11**(2), 105–113 (1996)
45. P. T. Treado, M. P. Nelson, *Raman Imaging*, vol. 2 (Wiley and Sons, Chichester, 2001), pp. 1429–1459
46. M.E. Auer, U.J. Griesser, J. Sawatzki, Qualitative and quantitative study of polymorphic forms in drug formulations by near infrared ft-raman spectroscopy. J. Mol. Struct. **661**, 307–317 (2003)
47. R. Weiss, M. Hessenberger, S. Kitzmuller, D. Bach, E.E. Weinberger, W.D. Krautgartner, C. Hauser-Kronberger, B. Malissen, C. Boehler, Y.N. Kalia, J. Thalhamer, S. Scheiblhofer, Transcutaneous vaccination via laser microporation. J. Controlled Release **162**(2), 391–399 (2012)
48. Y.G. Bachhav, A. Heinrich, Y.N. Kalia, Controlled intra- and transdermal protein delivery using a minimally invasive erbium:yag fractional laser ablation technology. Eur. J. Pharm. Biopharm. **84**(2), 355–364 (2013)
49. R.F. Donnelly, T.R.R. Singh, M.J. Garland, K. Migalska, R. Majithiya, C.M. McCrudden, P.L. Kole, T.M.T. Mahmood, H.O. McCarthy, A.D. Woolfson, Hydrogel-forming microneedle arrays for enhanced transdermal drug delivery. Adv. Funct. Mater. **22**(23), 4879–4890 (2012)
50. Yannic B. Schuetz, Aarti Naik, Richard H. Guy, Yogeshvar N. Kalia, Emerging strategies for the transdermal delivery of peptide and protein drugs. Expert opin. Drug Delivery **2**(3), 533–548 (2005)
51. T.M. Tuan-Mahmood, M.T.C. McCrudden, B.M. Torrisi, E. McAlister, M.J. Garland, T.R.R. Singh, R.F. Donnelly, Microneedles for intradermal and transdermal drug delivery. Eur. J. Pharm. Sci. **50**(5), 623–637 (2013)

52. H. Shapiro, L. Harris, F.W. Hetzel, D. Bar-Or, Laser assisted delivery of topical anesthesia for intramuscular needle insertion in adults. Lasers Surg. Med. **31**(4), 252–256 (2002)
53. J. Yu, Y.G. Bachhav, S. Summer, A. Heinrich, T. Bragagna, C. Bohler, Y.N. Kalia, Using controlled laser-microporation to increase transdermal delivery of prednisone. J. Controlled Release **148**(1), E71–E73 (2010)
54. S. Scheiblhofer, J. Thalhamer, R. Weiss, Laser microporation of the skin: prospects for painless application of protective and therapeutic vaccines. Expert Opin. Drug Delivery **10**(6), 761–773 (2013)
55. M. Haedersdal, F.H. Sakamoto, W.A. Farinelli, A.G. Doukas, J. Tam, R.R. Anderson, Fractional CO_2 laser-assisted drug delivery. Lasers Surg. Med. **42**(2), 113–122 (2010)
56. A. Vogel, V. Venugopalan, Mechanisms of pulsed laser ablation of biological tissues. Chem. Rev. **103**(2), 577–644 (2003)
57. X.H. Hu, Q.Y. Fang, M.J. Cariveau, X.N. Pan, G.W. Kalmus, Mechanism study of porcine skin ablation by nanosecond laser pulses at 1064, 532, 266, and 213 nm. IEEE J. Quant. Electron. **37**(3), 322–328 (2001)
58. J.S. Nelson, J.L. Mccullough, T.C. Glenn, W.H. Wright, L.H.L. Liaw, S.L. Jacques, Midinfrared laser ablation of stratum-corneum enhances in vitro percutaneous transport of drugs. J. Invest. Dermatol. **97**(5), 874–879 (1991)
59. C. Gomez, A. Costela, I. Garcia-Moreno, F. Llanes, J.M. Teijon, M.D. Blanco, Skin laser treatments enhancing transdermal delivery of ala. J. Pharm. Sci. **100**(1), 223–231 (2011)
60. R. Srinivasan, Ablation of polymers and biological tissue by ultraviolet-lasers. Science **234**(4776), 559–565 (1986)
61. R.R. Anderson, J.A. Parrish, Selective photothermolysis - precise microsurgery by selective absorption of pulsed radiation. Science **220**(4596), 524–527 (1983)
62. B. Brazzini, G. Hautmann, I. Ghersetich, J. Hercogova, T. Lotti, Laser tissue interaction in epidermal pigmented lesions. J. Eur. Acad. Dermatol. Venereol. **15**(5), 388–391 (2001)
63. G. Paltauf, P.E. Dyer, Photomechanical processes and effects in ablation. Chem. Rev. **103**(2), 487–518 (2003)
64. Y.G. Yingling, B.J. Garrison, Photochemical ablation of organic solids. Nucl. Instrum. Methods Phys. Res. Sect. B-Beam Interact. Mater. Atoms **202**, 188–194 (2003)
65. D.X. Hammer, R.J. Thomas, G.D. Noojin, B.A. Rockwell, P.K. Kennedy, W.P. Roach, Experimental investigation of ultrashort pulse laser-induced breakdown thresholds in aqueous media. IEEE J. Quantum Electron. **32**(4), 670–678 (1996)
66. A. Vogel, J. Noack, G. Huttman, G. Paltauf, Mechanisms of femtosecond laser nanosurgery of cells and tissues. Appl. Phys. B-Lasers Opt. **81**(8), 1015–1047 (2005)
67. P.K. Kennedy, A first-order model for computation of laser-induced breakdown thresholds in ocular and aqueous-media.1. theory. IEEE J. Quantum Electron. **31**(12), 2241–2249 (1995)
68. P.S. Tsai, P. Blinder, B.J. Migliori, J. Neev, Y.S. Jin, J.A. Squier, D. Kleinfeld, Plasma-mediated ablation: an optical tool for submicrometer surgery on neuronal and vascular systems. Curr. Opin. Biotechnol. **20**(1), 90–99 (2009)
69. A. Vogel, J. Noack, K. Nahen, D. Theisen, S. Busch, U. Parlitz, D.X. Hammer, G.D. Noojin, B.A. Rockwell, R. Birngruber, Energy balance of optical breakdown in water at nanosecond to femtosecond time scales. Appl. Phys. B-Lasers Opt. **68**(2), 271–280 (1999)
70. T.H. Maiman, Stimulated optical radiation in ruby. Nature **187**(4736), 493–494 (1960)
71. M.M. Zaret, G.M. Breinin, I.M. Siegel, H. Ripps, H. Schmidt, Ocular lesions produced by an optical maser (laser). Science **134**(348), 1525–1526 (1961)
72. L. Goldman, D.J. Blaney, D.J. Kindel, E.K. Franke, Effect of the laser beam on the skin. J. Invest. Dermatol. **40**(3), 121–122 (1963a)
73. L. Goldman, E.K. Franke, D.J. Kindel, D.J. Blaney, D. Richfield, Pathology of effect of laser beam on skin. Nature **197**(487), 912–914 (1963b)
74. L. Goldman, A. Freemond, P. Hornby, D.J. Blaney, Biomedical aspects of lasers. Jama-J. Am. Med. Assoc. **188**(3), 302–306 (1964)
75. L. Goldman, *Biomedical Aspects of the Laser: the Introduction of Laser Applications into Biology and Medicine* (Springer, Berlin, 1967)

76. S.L. Jacques, D.J. Mcauliffe, I.H. Blank, J.A. Parrish, Controlled removal of human stratum-corneum by pulsed laser. J. Invest. Dermatol. **88**(1), 88–93 (1987)
77. J.T. Walsh, T.J. Flotte, R.R. Anderson, T.F. Deutsch, Pulsed co_2-laser tissue ablation—effect of tissue-type and pulse duration on thermal-damage. Lasers Surg. Med. **8**(2), 108–118 (1988)
78. J.T. Walsh, T.J. Flotte, T.F. Deutsch, Er yag laser ablation of tissue—effect of pulse duration and tissue-type on thermal-damage. Lasers Surg. Med. **9**(4), 314–326 (1989)
79. W.R. Lee, S.C. Shen, K.H. Wang, C.H. Hu, J.Y. Fang, The effect of laser treatment on skin to enhance and control transdermal delivery of 5-fluorouracil. J. Pharm. Sci. **91**(7), 1613–1626 (2002)
80. A.J. Singer, R. Weeks, R. Regev, Laser-assisted anesthesia reduces the pain of venous cannulation in children and adults: a randomized controlled trial. Acad. Emerg. Med. **13**(6), 623–628 (2006)
81. C. Gomez, A. Costela, I. Garcia-Moreno, F. Llanes, J.M. Teijon, D. Blanco, Laser treatments on skin enhancing and controlling transdermal delivery of 5-fluorouracil. Lasers Surg. Med. **40**(1), 6–12 (2008)
82. B. Forster, A. Klein, R.M. Szeimies, T. Maisch, Penetration enhancement of two topical 5-aminolaevulinic acid formulations for photodynamic therapy by erbium:yag laser ablation of the stratum corneum: continuous versus fractional ablation. Exp. Dermatol. **19**(9), 806–812 (2010)
83. Y.G. Bachhav, S. Summer, A. Heinrich, T. Bragagna, C. Bohler, Y.N. Kalia, Effect of controlled laser microporation on drug transport kinetics into and across the skin. J. Controlled Release **146**(1), 31–36 (2010)
84. Y.G. Bachhav, A. Heinrich, Y.N. Kalia, Using laser microporation to improve transdermal delivery of diclofenac: increasing bioavailability and the range of therapeutic applications. Eur. J. Pharm. Biopharm. **78**(3), 408–414 (2011)
85. J. Yu, D.R. Kalaria, Y.N. Kalia, Erbium: yag fractional laser ablation for the percutaneous delivery of intact functional therapeutic antibodies. J. Controlled Release **156**(1), 53–59 (2011)
86. M. Hessenberger, R. Weiss, E.E. Weinberger, C. Boehler, J. Thalhamer, S. Scheiblhofer, Transcutaneous delivery of CPG-adjuvanted allergen via laser-generated micropores. Vaccine **31**(34), 3427–3434 (2013)
87. W.R. Lee, S.C. Shen, M.H. Pai, H.H. Yang, C.Y. Yuan, J.Y. Fang, Fractional laser as a tool to enhance the skin permeation of 5-aminolevulinic acid with minimal skin disruption: A comparison with conventional erbium:yag laser. J. Controlled Release **145**(2), 124–133 (2010)
88. C.S. Haak, B. Bhayana, W.A. Farinelli, R.R. Anderson, M. Haedersdal, The impact of treatment density and molecular weight for fractional laser-assisted drug delivery. J. Controlled Release **163**(3), 335–341 (2012)
89. X.Y. Chen, D. Shah, G. Kositratna, D. Manstein, R.R. Anderson, M.X. Wu, Facilitation of transcutaneous drug delivery and vaccine immunization by a safe laser technology. J. Controlled Release **159**(1), 43–51 (2012)
90. J. Lippert, R. Smucler, M. Vlk, Fractional carbon dioxide laser improves nodular basal cell carcinoma treatment with photodynamic therapy with methyl 5-aminolevulinate. Dermatol. Surg. **39**(8), 1202–1208 (2013)
91. G. Nicolodelli, D. P. Angarita, N. M. Inada, L. F. Tirapelli, V. S. Bagnato. Effect of photodynamic therapy on the skin using the ultrashort laser ablation. J. Biophotonics (2013)
92. P.A. Todd, K.L. Goa, Interferon gamma-1b—a review of its pharmacology and therapeutic potential in chronic granulomatous-disease. Drugs **43**(1), 111–122 (1992)
93. M.H. Tan, J.S. Dover, T.S. Hsu, K.A. Arndt, B. Stewart, Clinical evaluation of enhanced nonablative skin rejuvenation using a combination of a 532 and a 1,064 nm laser. Lasers Surg. Med. **34**(5), 439–445 (2004)
94. C.H. Lin, I.A. Aljuffali, J.Y. Fang, Lasers as an approach for promoting drug delivery via skin. Expert Opin. Drug Delivery **11**(4), 599–614 (2014)
95. D. Manstein, G.S. Herron, R.K. Sink, H. Tanner, R.R. Anderson, Fractional photothermolysis: a new concept for cutaneous remodeling using microscopic patterns of thermal injury. Lasers Surg. Med. **34**(5), 426–438 (2004)

96. Clinical results—P.L.E.A.S.E. professional (2014)
97. U. Paasch, M. Haedersdal, Laser systems for ablative fractional resurfacing. Expert Rev. Med. Devices **8**(1), 67–83 (2011)
98. E.M. Graber, E.L. Tanzi, T.S. Alster, Side effects and complications of fractional laser photothermolysis: experience with 961 treatments. Dermatol. Surg. **34**(3), 301–307 (2008)
99. L.R. Sklar, C.T. Burnett, J.S. Waibel, R.L. Moy, D.M. Ozog, Laser assisted drug delivery: a review of an evolving technology. Lasers Surg. Med. **46**(4), 249–262 (2014)
100. J. Neev, L.B. DaSilva, M.D. Feit, M.D. Perry, A.M. Rubenchik, B.C. Stuart, Ultrashort pulse lasers for hard tissue ablation. IEEE J. Sel. Top. Quantum Electron. **2**(4), 790–800 (1996)
101. K.M. Kent, E.M. Graber, Laser tattoo removal: a review. Dermatol. Surg. **38**(1), 1–13 (2012)
102. T. Juhasz, H. Frieder, R.M. Kurtz, C. Horvath, J.F. Bille, G. Mourou, Corneal refractive surgery with femtosecond lasers. IEEE J. Sel. Top. Quantum Electron. **5**(4), 902–910 (1999)
103. M. Dutra-Correa, G. Nicolodelli, J.R. Rodrigues, C. Kurachi, V.S. Bagnato, Femtosecond laser ablation on dental hard tissues-analysis of ablated profile near an interface using local effective intensity. Laser Phys. **21**(5), 965–971 (2011)
104. K.S. Frederickson, W.E. White, R.G. Wheeland, D.R. Slaughter, Precise ablation of skin with reduced collateral damage using the femtosecond-pulsed, terawatt titanium-sapphire laser. Arch. Dermatol. **129**(8), 989–993 (1993)
105. H. Huang, Z. X. Guo, Human dermis separation via ultra-short pulsed laser plasma-mediated ablation. J. Phys. D-Appl. Phys. **42**(16) (2009)
106. N. Suhm, M.H. Gotz, J.P. Fischer, F. Loesel, W. Schlegel, V. Sturm, J. Bille, R. Schroder, Ablation of neural tissue by short-pulsed lasers—a technical report. Acta Neurochir. **138**(3), 346–349 (1996)
107. A.A. Oraevsky, L.B. DaSilva, A.M. Rubenchik, M.D. Feit, M.E. Glinsky, M.D. Perry, B.M. Mammini, W. Small, B.C. Stuart, Plasma mediated ablation of biological tissues with nanosecond-to-femtosecond laser pulses: relative role of linear and nonlinear absorption. IEEE J. Sel. Top. Quantum Electron. **2**(4), 801–809 (1996)
108. C.C. Sumian, F.B. Pitre, B.E. Gauthier, M. Bouclier, S.R. Mordon, Laser skin resurfacing using a frequency doubled nd : yag laser after topical application of an exogenous chromophore. Lasers Surg. Med. **25**(1), 43–50 (1999)
109. J.Y. Jung, J.S. Hong, C.H. Ahn, J.Y. Yoon, H.H. Kwon, D.H. Suh, Prospective randomized controlled clinical and histopathological study of acne vulgaris treated with dual mode of quasi-long pulse and q-switched 1064-nm nd:yag laser assisted with a topically applied carbon suspension. J. Am. Acad. Dermatol. **66**(4), 626–633 (2012)
110. M.R. Roh, H.J. Chung, K.Y. Chung, Effects of various parameters of the 1064 nm nd:yag laser for the treatment of enlarged facial pores. J. Dermatol. Treat. **20**(4), 223–228 (2009)

Chapter 3
Methodology

3.1 Film-Forming Systems

3.1.1 FFS Preparation

All materials used to fabricate polymeric film-forming systems (FFS) were kindly provided by LEO Pharma, Denmark. The polymers used were Eudragit® RS PO (Eudragit) (ammonio methacrylate copolymer type B) and Klucel™ LF (Klucel) (hydroxypropyl cellulose). Additives such as triethyl citrate (TEC) and medium-chain triglyceride (MCT) (Miglyol 812 N, caprylic/capric triglyercide) and the model drug betamethasone-17-valerate (BMV, purity 100 %) were also incorporated.

FFS were prepared by combining a polymer and absolute ethanol (mixed with 5 % w/w water for Eudragit) (percentage contents are always reported as mass fractions hereinafter). For the incorporation of TEC or MCT, 10 % stocks of TEC/MCT in absolute ethanol were prepared by stirring for 2 h. TEC/MCT stock was mixed with the appropriate amount of polymer to achieve concentrations of 20 or 40 % (percentages of TEC and MCT correspond to the percentage of dry polymer weight). The solutions were stirred overnight to ensure complete dissolution of the polymer. A clear solution was obtained after stirring. Drug loaded FFS were prepared by dissolving 1.2 % BMV (corresponding to 1.0 % betamethasone) in the FFS. Table 3.1 summarizes the compositions of the FFS investigated in this work.

3.1.2 Atomic Force Microscopy Imaging

AFM experiments were performed in ambient conditions using a Multimode Scanning Probe Microscope (Veeco, Plainview, NY) with a Nanoscope IIIA controller and Nanoscope software (Version 7.341).

H. Garvie-Cook, *Novel (Trans)dermal Drug Delivery Strategies*,
Springer Theses, DOI 10.1007/978-3-319-28901-4_3

Table 3.1 Composition of the FFS tested

Formulation	Percentage by weight (%)			
	Polymer	TEC/MCT	Absolute ethanol	Water
15 % Eudragit without plasticizer	15	0	80	5
15 % Eudragit with 20 % TEC	15	3	77	5
15 % Eudragit with 40 % TEC	15	6	74	5
15 % Eudragit with 20 % MCT	15	3	77	5
5 % Klucel without plasticizer	5	0	95	0
5 % Klucel with 20 % TEC	5	1	94	0
5 % Klucel with 40 % TEC	5	2	93	0
5 % Klucel with 20 % MCT	5	1	94	0

3.1.2.1 Film Formation

For AFM imaging and nanoindentation, 10 μl cm^{-2} of the FFS were cast uniformly onto glass slides, which had been cleaned in an ultrasonic bath with acetone and then isopropanol for ten minutes in each. Unless otherwise stated, glass slides were maintained overnight at a temperature of approximately 30 °C, to mimic the temperature of the outer surface of the skin [1]. Formed films had thicknesses of approximately 10 μm.

Films were additionally formed on porcine skin for some nanoindentation investigations by casting 10 μl cm^{-2} of FFS on their surface. Details of the preparation of the skin are given in Sect. 3.2.1. The skin was kept moist by applying a thin layer of silicone grease (Rhodorsil, Hong Kong) between the bottom of the skin and the AFM stub on which is was supported. Films on skin were dried for 6 h at room temperature.

3.1.2.2 Image Acquisition

AFM images were obtained in tapping mode in ambient conditions. 'All in One' AFM probes (AlOAl, Budget Sensors, Sofia, Bulgaria), with nominal spring constants of between 0.2 and 40 N m^{-1} and resonance frequencies between 15 and 350 kHz, were used for both imaging and nanoindentation. Images were analysed using Nanoscope Analysis software (Version 1.3, Bruker, Billerica, MA). The frequency of oscillation used for imaging was generally determined using the autotune function in the Nanoscope controller software. A scanning electron microscopy (SEM) image of an AFM cantilever is shown in Fig. 3.1a.

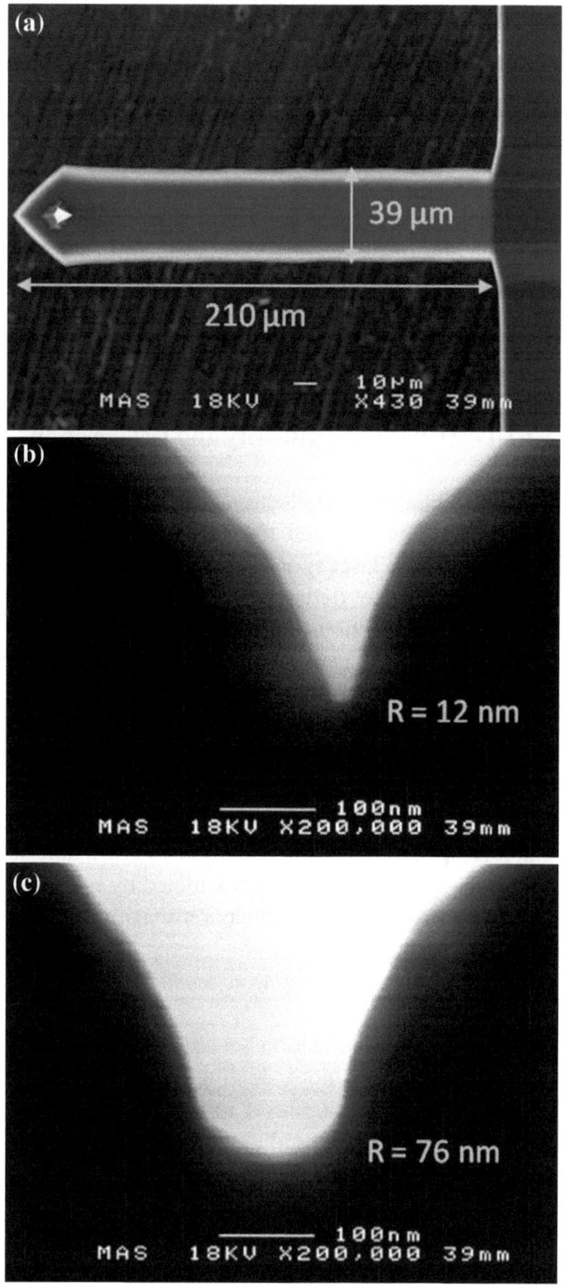

Fig. 3.1 Example SEM images of AFM probes. Measurements of **a** cantilever length and width, and tip radius of curvature **b** before and **c** after EBID are shown in *yellow*

3.1.3 AFM Nanoindentation

The process of AFM nanoindentation of polymeric films, prepared according to Sect. 3.1.2.1, and a typical cantilever deflection-displacement curve are summarized in Fig. 3.2a. AFM nanoindentation was carried out in contact mode (non-oscillating cantilever).

The raw data acquired in AFM nanoindentation were cantilever deflection (in volts) as a function of the vertical displacement of the cantilever from its initial position above the sample surface. There were several steps to convert these data to a form which could be used to determine the type of deformation observed (be it viscous, elastic and/or plastic) and to extract the elastic modulus.

3.1.3.1 Deflection Sensitivity and Deformation

The AFM cantilever deflection was detected using the position of a laser beam, which had been reflected from the back of the cantilever, on to a four-quadrant photodiode. The laser spot was initially positioned in the middle of the photodiode and any subsequent deflection of the cantilever was detected as a change in output voltage. The deflection of the cantilever was converted from voltage to distance by determining the deflection sensitivity of the cantilever, which is the distance that the cantilever deflects per volt measured by the photodiode. The deflection sensitivity was determined by calibrating the cantilever on a surface which deformed negligibly under the applied load (cleaned glass slides were used). The deflection sensitivity is the inverse of the gradient of the cantilever deflection (in Volts) as a function of its vertical displacement (in nanometers) (i.e., the cantilever deflects the same distance that it is pushed towards the negligibly deformable surface). The deflection D of the cantilever in nanometers was calculated by multiplying the deflection, as measured by the photodiode, by the deflection sensitivity,

$$D(\mathrm{nm}) = \mathit{sensitivity}(\mathrm{nm\ V^{-1}}) \times D(\mathrm{V}) \tag{3.1}$$

An example of a deflection sensitivity calibration is shown in Fig. 3.2a (dashed line). The probe was not indenting the sample and the response observed is merely the response of the cantilever to being pushed onto the sample. Typical values of deflection sensitivities used in this work ranged from 36.5 to 68.6 nm V^{-1}.

The deformation of the sample (the depth of the indent) at a given load is the difference in vertical position of the cantilever between the sensitivity calibration and the experimental data (labelled in Fig. 3.2a).

3.1.3.2 Load

Once the deflection of the cantilever had been converted from volts to nanometers, the load P on the sample during indentation was determined using the cantilever's

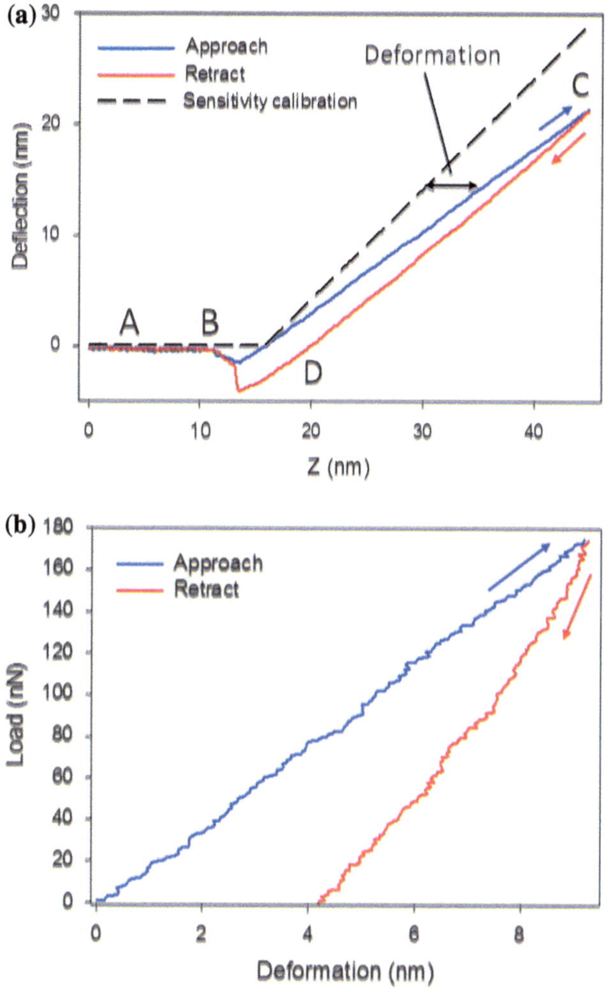

Fig. 3.2 **a** An example deflection-displacement curve obtained in the AFM nanoindentation of Eudragit films without plasticizer. The deflection of the AFM cantilever as a function of its vertical displacement from its original position was measured. *A* The probe tip is out of contact with the surface of the sample. The cantilever is brought down towards the sample surface at a constant approach rate. *B* The probe tip snaps into contact with the sample due to attractive van der Waals forces. At this point, the cantilever is bent towards the sample (negative deflection). As the approach continues, the cantilever bends upwards and becomes flat again when its deflection is zero. *C* While the tip is pushed into the sample, the cantilever continues to bend upwards until reaching a maximum deflection of 21 nm. *D* The cantilever is retracted and its deflection decreases. Adhesion between the probe tip and the sample surface causes the deflection to become negative while retracting. The tip then breaks free of this adhesion and the cantilever continues its ascent undeflected. **b** Deformation of the sample with load applied (calculated from data in **a**). The deformation of the sample is the difference between the sensitivity calibration, on a negligibly deformable surface, and the approach/retract data at a given load

spring constant k_C,

$$P = k_C D \tag{3.2}$$

k_C was calculated using a previously published method [2], requiring knowledge of the density of the medium (in this case, air) in which the cantilever is oscillated ρ_m, the width b and length L of the cantilever, the quality factor of the fundamental mode of the cantilever in the medium Q_m, and the resonant frequency in the medium ω_m:

$$k_C = 0.1906 \rho_m b^2 L Q_m \Gamma_i(\omega_m) \omega_m^2 \tag{3.3}$$

where Γ_i is the imaginary component of the hydrodynamic function $\Gamma(\omega)$, described by Eq. 20 in [3]. This function only depends on the Reynolds number,

$$Re = \rho_f \omega b^2 / 4\eta \tag{3.4}$$

where η is the viscosity of the medium. The density and viscosity of air are 1.18 kg m^{-3} and 1.86×10^5 kg m^{-1} s^{-1} respectively. The width and length of the cantilever are required for spring constant determination and were measured using scanning electron microscopy (SEM) (see Sect. 3.1.3.3). The quality factor and resonant frequency were measured using the autotune function in the AFM Nanoscope software. Spring constants between 1.7 and 36.1 N m^{-1} were used in these experiments.

For interpretation of deformation behaviour, it is common to show the deformation of the sample with the applied load (hereinafter referred to as load-deformation plots). Expressing the data in this way also allows a model to be fitted and the elastic modulus to be extracted. This is further discussed in Sect. 3.1.3.4. An example of a load-deformation plot, converted from cantilever deflection as a function of displacement, is shown in Fig. 3.2b.

3.1.3.3 Scanning Electron Microscopy of AFM Probes

The dimensions of the AFM probe cantilever and the shape of the probe tip (the indenter) were imaged using a Field Emission Scanning Electron Microscope (FESEM) (6301F, JEOL, Tokyo, Japan) and were measured with Image J (1.47v). The shape of the indenter is required for the extraction of elastic modulus from indentation data (see Sect. 3.1.3.4). Images of probe tips and cantilever lengths and widths were taken using an accelerating voltage of 18 kV. Example images are shown in Fig. 3.1.

Probe tips of rounded geometry were required for nanoindentation of polymeric films. Indentation is commonly carried out using flat punch, spherical or conical (or pyramidal) indenters [4]. The radius of curvature of AFM probe tips as manufactured can be as small as a few nanometres. These sharp tips apply a large pressure over their small contact area with the sample. For soft materials, such as polymers and biological samples, the use of spherical indenters, with larger radii than typical

AFM tips, results in a lower contact pressure and therefore less plastic deformation. This allows elastic information to be extracted more reliably [5–7].

The use of a larger tip for indentation does, however, limit the resolution of the technique. Spherical indenters are commonly fabricated by attaching, for example, glass spheres to AFM cantilevers which do not have tips [5]. The diameters of these spheres start from approximately 0.5 μm and the practical difficulties of attaching the spheres to the cantilever increase as the diameter of the spheres decrease. To fabricate spherical indenters with radii in the range of tens of nanometres, electron beam induced deposition (EBID) can be used.

EBID arises from the decomposition of hydrocarbon molecules present as contamination in SEM. By focussing the electron beam on an AFM probe tip, amorphous carbon is deposited and the thickness of the tip increases. For further description of EBID of AFM probes, see [8, 9]. A lower accelerating voltage is used for EBID of AFM probe tips than for imaging. Under these conditions, the interaction volume of the incident electrons with the atoms of the sample remains optimally within the AFM probe [10]. If a higher acceleration voltage is used, the electron beam mostly passes through the tip and little EBID occurs.

For indentation, the radius of AFM probe tip was increased in an SEM by bringing it into the electron beam focus at an accelerating voltage of 10 kV. The beam was kept on the tip at high magnification (×200000) for 30–60 min. The deposition process was stopped when the radius of the probe tip was sufficiently large. An example of an AFM probe tip before and after EBID is shown in Fig. 3.1. The maximum depths of the indentation carried out in this work were less than the radii of curvature of the tips, allowing the indenter to be modelled as spherical.

3.1.3.4 Elastic Modulus Extraction

The model used to extract material properties of a sample under indentation must be carefully selected to comply with the kinds of deformation observed. Fingerprint features in load-deformation plots were used to identify the kinds of deformation occurring [7]. In the indentation of polymeric films, viscous, elastic and plastic deformation was observed. This is further discussed in Sect. 4.2.1.

The elastic contact between two spherical surfaces with different radii and elastic moduli was originally considered by Hertz in 1882 and was developed by Johnson for spherical indentation [11]. For this model to apply, there must be no additional interactions between the indenter and sample (such as adhesion) [5]. For purely elastic samples, the relation between the load from a spherical indenter and the deformation h it causes is given by:

$$P = \frac{4\sqrt{R}}{3} E_r h^{\frac{3}{2}} \tag{3.5}$$

where R is the radius of the indenter and E_r is the reduced elastic modulus of the sample. Reduced modulus takes into account the elastic moduli of the sample E and the indenter E_i,

$$\frac{1}{E_r} = \frac{(1 - \nu_i^2)}{E_i} + \frac{(1 - \nu^2)}{E} \tag{3.6}$$

where ν_i and ν are the Poisson's ratios of the indenter and of the sample, respectively. When the indenter modulus is much greater than the sample modulus (in this work, 150 GPa for silicon [12] and 28 GPa for EBID [8]), the elastic modulus of the sample can be approximated as,

$$E = E_r(1 - \nu^2) \tag{3.7}$$

A value of 0.4 was used for the Poisson's ratios of Eudragit and Klucel films. This approximate value has previously been found for polymeric films of a similar nature [13].

In purely elastic samples, the approach and retract responses of the sample to indentation are the same (no hysteresis), so the model can be fitted to either stage (loading or unloading) of the process. If viscous and/or plastic deformation is apparent, the initial stages of the loading process during indentation can be modelled as elastic. Plastic deformation is characterized by a threshold. The deformation of the sample below this threshold (in an elastic-plastic sample) is therefore purely elastic and can be modelled using the Hertzian approach. Viscous deformation has no threshold for onset but is time-dependent. Using data from the initial loading period will therefore allow little time for viscous flow to affect the response of the sample and the elastic properties can be extracted [7].

Instead of assuming the limit of purely elastic deformation during indentation of viscous-elastic-plastic samples, a model was developed by Oyen & Cook to account for plastic deformation and for the time-dependent viscous deformation [4]. This model was based on the series elastic-viscous Maxwell model and was extended to include an element for plasticity. These elements are depicted in Fig. 3.3.

The Oyen & Cook model was initially developed for indentation using a flat punch. The area of contact between an indenter of this geometry and the sample remains constant throughout indentation. The elastic deformation of the sample is proportional to the load (Fig. 3.3), as opposed to the power of 3/2 found for spherical

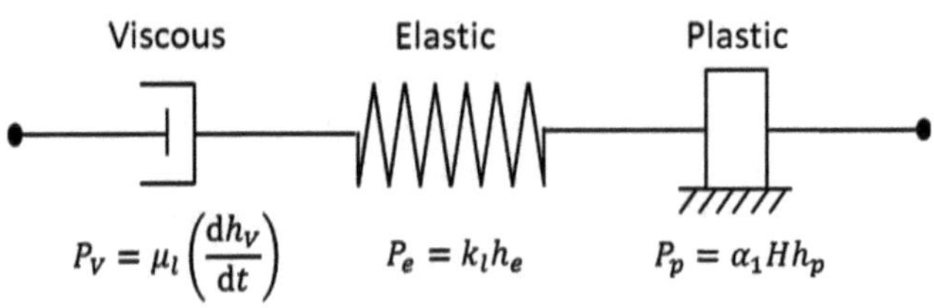

Fig. 3.3 Deformation element series model for viscous-elastic-plastic indentation. Relationships between load P and deformation h, for indentation using a flat punch, are shown for the viscous element (P_V), elastic spring (P_e) and an element for plasticity (P_p), where μ_l is the linear viscosity coefficient, t is the time during indentation, k_l is the elastic spring constant, α_1 is a coefficient based on the geometry of the indenter, and H is the hardness

indentation. The load on the viscous element is proportional to the deformation rate $\frac{dh_V}{dt}$, where h_V is viscous deformation and t is time. Oyen & Cook extended this model to indentation using a conical or pyramidal indenter, where the elastic load P_e and deformation h_e have a quadratic relationship,

$$P_e = k_Q h_e^2 \tag{3.8}$$

where k_Q is the quadratic stiffness related to the elastic modulus of the sample. When indenting using a flat punch, the same power law applies for elastic and viscous deformation. Oyen & Cook therefore assumed that the same power law for elastic deformation using a conical indenter could be applied to viscous and plastic deformation using a conical indenter. The same approach is used here to develop a model for viscous-elastic-plastic indentation with a spherical indenter. The power law for elastic load-deformation during indentation using a spherical punch is 3/2 (Eq. 3.5). The relations for time-dependent and plastic deformation are therefore raised to the same power. The load on the viscous element P_V when indented using a spherical indenter is:

$$P_V = \mu \left(\frac{\mathrm{d}h_V}{\mathrm{d}t} \right)^{\frac{3}{2}} \tag{3.9}$$

where μ is a viscosity coefficient and t is the time during indentation. The load P_p on the plastic element of deformation is given by:

$$P_p = \alpha H h_p^{\frac{3}{2}} \tag{3.10}$$

where h_p is plastic deformation, α is related to the geometry of the indenter and H is the hardness.

The total deformation of the sample is determined from the series combination of the individual elements:

$$h = h_V + h_e + h_p \tag{3.11}$$

and the deformation rate follows from this, i.e.,

$$\frac{\mathrm{d}h}{\mathrm{d}t} = \frac{\mathrm{d}h_V}{\mathrm{d}t} + \frac{\mathrm{d}h_e}{\mathrm{d}t} + \frac{\mathrm{d}h_p}{\mathrm{d}t} \tag{3.12}$$

The total load is the same in the viscous, elastic and plastic elements, $P = P_V = P_e = P_p$. Substituting Eqs. 3.5, 3.9 and 3.10 into Eq. 3.12 yields an expression relating the deformation rate to the applied load:

$$\frac{\mathrm{d}h}{\mathrm{d}t} = \left(\frac{P}{\mu} \right)^{\frac{2}{3}} + \frac{2}{3} \left(\frac{1}{k^2 P} \right)^{\frac{1}{3}} \frac{\mathrm{d}P}{\mathrm{d}t} + \frac{2}{3} \left(\frac{1}{\alpha^2 H^2 P} \right)^{\frac{1}{3}} \frac{\mathrm{d}P}{\mathrm{d}t} \tag{3.13}$$

where $k = \frac{4\sqrt{R}}{3}\frac{E}{(1-\nu^2)}$. This relation can be solved for a variety of loading conditions, such as different approach rates and the use of a delay at the maximum load, when the probe is held in contact with the sample. Loading conditions can be varied to observe the change in the relative contributions of elastic, plastic and viscous deformations.

In these experiments, trapezoidal loading conditions (with a hold) and loading conditions without a hold were used. Trapezoidal loading conditions are shown in Fig. 3.4. The load on the sample during the three (or two, without a hold) stages of the indentation process were:

$$P = \frac{P_{max}}{\tau_L} t \quad \text{when} \quad 0 \le t \le \tau_L$$

$$P = P_{max} \quad \text{when} \quad \tau_L \le t \le \tau_L + \tau_H$$

$$P = \frac{P_{max}}{\tau_U}(\tau_L + \tau_H + \tau_U - t) \quad \text{when} \quad \tau_L + \tau_H \le t \le \tau_L + \tau_H + \tau_U$$

where P_{max} is the maximum load applied to the sample, τ_L is the duration of loading, τ_H is the hold period, and τ_U is the duration of unloading.

While increasing the load on the sample at a constant rate, elastic, plastic and viscous deformation occur. During loading, the relation between the total deformation

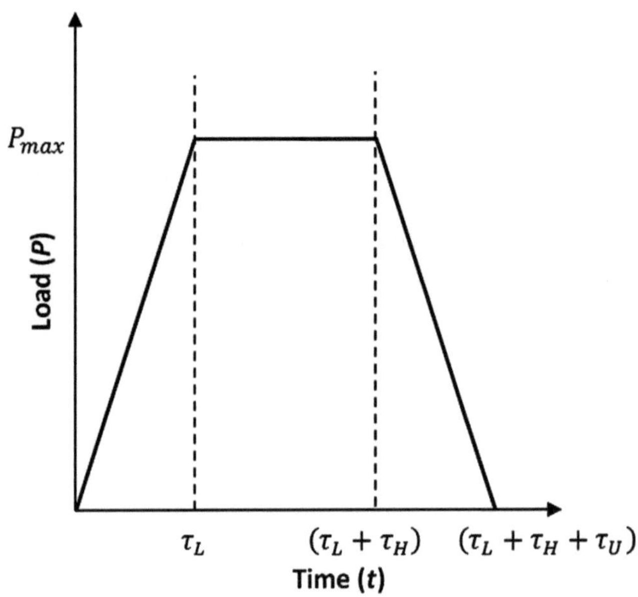

Fig. 3.4 Loading conditions used in AFM indentation experiments. The probe approaches the sample at a constant rate for a duration τ_L until a maximum load P_{max} has been reached. The AFM cantilever can then be held at its maximum vertical displacement from its initial position, for a duration τ_H. The probe is then retracted at a constant unloading rate for a duration τ_U until it is finally out of contact

and the load applied therefore includes elastic, viscous and plastic terms. When a constant load is applied, only viscous deformation occurs.

It is usually assumed that all plastic deformation occurs during the loading stage of the indentation process [4]. During unloading, therefore, $\frac{dh_p}{dt}$ is zero and the relation between deformation rate and applied load reduces to:

$$\frac{dh}{dt} = \left(\frac{P_{max}}{\mu\tau_U}\right)^{\frac{2}{3}} (\tau_L+\tau_H+\tau_U-t)^{\frac{2}{3}} - \frac{2}{3}\left(\frac{P_{max}}{k\tau_U}\right)^{\frac{2}{3}} (\tau_L+\tau_H+\tau_U-t)^{-\frac{1}{3}} \quad (3.14)$$

Integrating this with respect to time yields the following expression for deformation during unloading:

$$h(t) = \left(\frac{P_{max}}{\tau_U}\right)^{\frac{2}{3}} (\tau_L+\tau_H+\tau_U-t)^{\frac{2}{3}} \left[\frac{1}{k^{\frac{2}{3}}} - \frac{3}{5}\frac{1}{\mu^{\frac{2}{3}}}(\tau_L+\tau_H+\tau_U-t)\right] + h_f \quad (3.15)$$

where h_f is the residual deformation left after all load has been removed.

The Hertzian and Oyen & Cook models were used for the analysis of the viscous-elastic-plastic indentation of polymeric films. To extract elastic modulus, data were analysed in the form of load-deformation plots for the Hertzian model fit and deformation-time plots for the Oyen & Cook model fit. The region of the initial loading stage of the indentation that was used to fit the Hertzian model was the first 5–20 % of data points. This range excluded the noisiest part of the data at low load values (first 5 %) [14]. The fit range included the next 15 % of data points to avoid taking into account viscous and plastic deformation. The Oyen & Cook model was fit to the middle 75 % of data. This was again to avoid the noisiest parts of the data. Unless otherwise stated, indents were performed using an approach rate of 16 nm s^{-1} and a surface hold of 0 s.

3.1.4 Raman Micro-Spectroscopy

Spectroscopic measurements were performed using a Raman microscope (inVia, Renishaw, Gloucestershire, UK) which measures the Stokes shift of scattering substances. To obtain Raman spectra of polymeric films and their constituents, an excitation wavelength of 532 nm was employed.

Initially, Raman spectra of the constituents of polymeric films were acquired using 10 % of the maximum laser power available (approximately 80 mW). Spectra of Eudragit, Klucel, TEC and MCT in the solid state and in solution were obtained. BMV was prepared in three different ways to determine the relation between the peak position and the physical state of the drug. BMV was investigated in the crystalline state, as provided by the supplier, in the amorphous state, formed by depositing a drop of an ethanolic solution onto a cleaned glass slide [15], and in solution (0.4 % in ethanol). Characteristic peaks apparent in the spectra of TEC, MCT and

BMV, which were distinct from spectral features of the polymers, permitted their spatial distribution to be mapped.

Raman maps of Eudragit and Klucel films incorporating 20 % TEC or MCT were acquired using a ×50 objective. To increase resolution, the excitation laser beam was directed to the sample via a pinhole. This reduced the area from which spectra were obtained, and therefore the area of each individual pixel of the map, to $1 \times 1\ \mu m^2$. Mapping was performed using the maximum laser power available and no damage to any sample was observed.

Raman spectra were analysed using Wire software (3.4 and 4.0, Renishaw). Background signals were subtracted using the Cubic Spine Interpolation method with the baseline either defined by the software or the user. Spectral parameters were defined and fitted to specific peaks for the identification of TEC, MCT and BMV. The chemical compositions of the films were mapped using the intensities of fitted peaks and also via the intensities corresponding to specific spectral shifts (if a distinct peak could not be fitted). The physical state of BMV within the films was mapped using the position of its characteristic peak and its relation to calibration spectra for crystalline, amorphous and dissolved BMV.

3.2 Laser Microporation

3.2.1 Preparation of Skin Samples

Porcine dorsal skin was used for laser microporation and AFM indentation (see Sect. 3.1.3). Porcine skin is a recognised and representative model for human skin [16] and is readily available and inexpensive compared to human tissue. The skin was obtained from a local slaughterhouse shortly after sacrifice of the animal, and was subsequently cleaned in cold water and dermatomed (Zimmer, Warsaw, IN) to approximately 750 μm thick. The skin sections were then individually wrapped in ParafilmTM and kept in a freezer at −20 °C. Shortly before use, the skin was thawed and excess hair was carefully removed using scissors.

For AFM experiments, skin was cut into pieces of approximately $1 \times 1\ cm^2$. Samples for release experiments were cut to a larger area to ensure that they completely covered the area of contact between the donor and acceptor chambers of the Franz diffusion cells (for details, see Sect. 3.2.5). For laser microporation, skin samples were placed on glass slides and supported in place using Parafilm™.

The skin was either prepared as above or ink was applied to its surface (now referred to as uninked or inked skin, respectively). A black permanent marker (Lyreco, Shropshire, UK) was used for this purpose and was drawn across the surface of the sample. For release experiments on multiple pieces of skin, a consistent area of ink was obtained using a stencil.

To determine a method with which to characterise the extent of thermal damage, porcine skin was purposefully burnt using a soldering iron. Skin was burnt at temperatures of 420, 160 and 60 °C. The highest temperature was obtained by contacting the skin directly with the soldering iron. The contact was maintained until the temperature, measured by a thermocouple in between the skin and the soldering iron, reached 420 °C. Lower temperatures were achieved by diffusing the heat using a copper washer between the soldering iron and the skin. Again, the heating source was removed when the temperature had reached the desired level. The dependence of Raman scattering from the areas of the skin burnt at different temperatures was then acquired (see Sect. 3.2.4).

3.2.2 Laser Set-Up

A Ytterbium doped fibre laser (Fianium, Southampton, UK), with a wavelength of 1064 nm, pulse duration of 5 ps and a repetition rate of 20 MHz was used in these experiments. Pulses were compressed using a grating compressor pair and frequency doubled in a 3 mm lithium triborate (LBO) crystal. The laser beam resulting from this set-up, which was used to ablate porcine skin, had a wavelength of 532 nm and pulse duration of approximately 300 fs. A camera shutter was used to expose the skin to a range of exposure times.

The laser was either focussed on the skin using a lens (magnification of ×20, numerical aperture (NA) of 0.4) or directed towards the skin using a hollow core photonic crystal fibre (PCF). The skin was mounted on a translatable stage, with its surface perpendicular to the propagation of the laser beam. In the lens set-up, the optimum focus was found by continually exposing the skin to the laser beam and adjusting the horizontal distance between the skin and the objective until a strong effect was observed (smoking from the skin, flash of light).

The fibre was set up relative to the surface of the skin using an optical microscope which was positioned above the equipment. The hollow core PCF used was designed to guide light with a wavelength around 532 nm. The experimental set up involving the use of fibres to guide light to the skin is shown in Fig. 3.5a. An SEM image of a fibre of the same structure of that used (though a larger diameter) is shown in Fig. 3.5b. The diameter of the core of the fibre was 16 μm and the NA was approximately 0.04.

3.2.3 Optical Microscopy

An optical microscope was used to obtain images and measure the dimensions of laser micropores produced in porcine skin. A calibration of the number of pixels per millimetre, depending on the objective used, was initially obtained using a grid of lines of known separation. The diameter of the pores in the optical microscope images could therefore be measured in Image J software and converted to the actual length using this calibration.

(a)

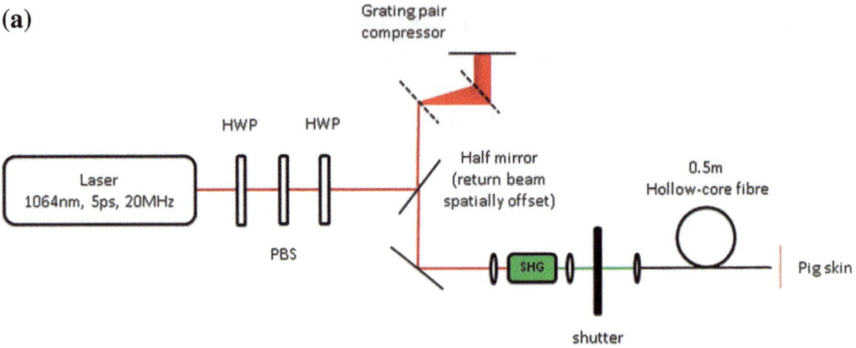

(b)

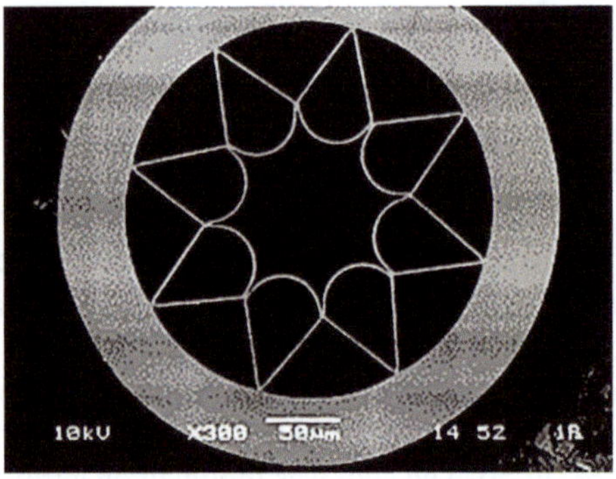

Fig. 3.5 **a** Doubling the frequency and compressing the pulses of a fibre laser for ablation of porcine skin; **b** SEM image of a hollow core PCF with the same structure as, but a larger diameter than, the fibre used in these experiments. Images are reproduced by kind permission of James Stone and Fei Yu, respectively, both from the University of Bath

Depth was calibrated by determining the displacement of the dial which controlled the stage height of the microscope required to adjust the focus. The focus was initially on the top surface of a glass slide. Once the glass slide had been removed, the height dial was then moved so that the stage, which had previously been underneath the glass slide, was in focus. The thickness of the slide was measured using a micrometer and the height dial displacement was then calibrated. The depths of laser micropores in skin were measured by initially focussing the microscope on the top surface of the skin (the rim of the pores) then moving the height dial until the bottom of the pores was in focus. The depth was determined as the difference in height between the initial and final focusses.

3.2.4 Raman Micro-Spectroscopy of Porcine Skin

Raman micro-spectroscopy of porcine skin (both with and without laser micropores) was performed using the same device as previously described (Sect. 3.1.4). An

Table 3.2 Exposure times and laser powers used to acquire Raman spectra of undamaged and burnt porcine skin

Extent of damage	Exposure time (s)	Laser power (% of maximum)
Unburnt	12	100
Burnt to 60 °C	12	100
Burnt to 160 °C	12	10
Burnt to 420 °C	12	0.0001

excitation wavelength of 785 nm was used. Spectra of porcine skin obtained using this wavelength showed less fluorescent background and revealed more spectral features than those obtained with a wavelength of 532 nm.

Spectra of skin burnt at temperatures of 420, 160 and 60 °C and unburnt skin (see Sect. 3.2.1) were acquired using a ×20 objective. The power used for acquisition of Raman spectra varied according to the extent of the burn. The fluorescence greatly increased as the temperature of the burn increased, resulting in the use of lower laser powers. Spectra were then normalised according to power. The settings used to obtain Raman spectra of the above mentioned skin samples are shown in Table 3.2. The process was carefully monitored to ensure that the skin was not damaged further by the excitation laser beam.

Raman spectra of laser micropores were acquired to determine the extent of thermal damage to the surrounding tissue. Initially, a spectrum was acquired of inked skin that had not been porated using an exposure time of 12 s, 0.05 % of the maximum laser power available, and a ×20 objective. Spectra of laser micropores were acquired using a ×50 objective. Spectra were acquired from the bottom of laser micropores by focussing the Raman microscope on the bottom of the pore and by placing a pinhole in the path of the excitation beam, limiting the volume of its interaction with the sample. Laser powers of 0.5 and 100 % of the maximum power were used to obtain spectra of the bottom of micropores in uninked and inked skin, respectively.

3.2.5 *Caffeine Delivery*

Arrays of 18 × 18 micropores were produced on inked skin using two laser powers: 0.06 and 0.30 W. Inked skin was mounted on a translatable stage, perpendicular to the fibre used to direct the laser light, and was moved approximately 130 μm between 1 s exposures. The skin was positioned such that the most protruding surface of the skin was at a distance of approximately 300 μm from the end of the fibre.

Skin samples for delivery experiments, across control and porated barriers, were thawed and had ink applied. Producing an 18 × 18 array of micropores took approximately 40 min and 2 samples of skin were porated on each day that the experiments were performed.

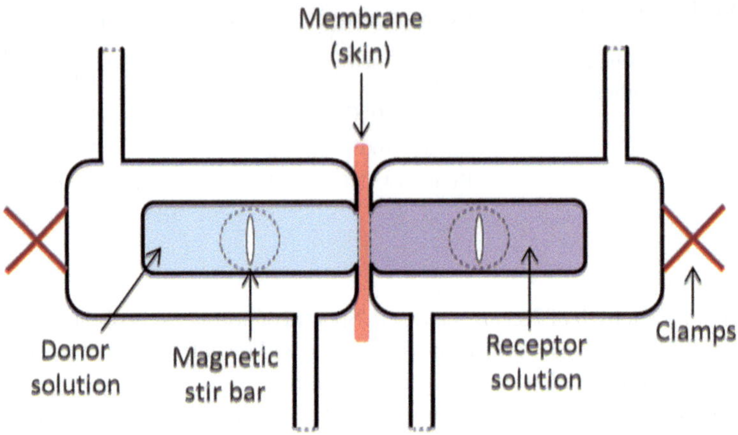

Fig. 3.6 Diagram of a side-by-side Franz diffusion cell

After poration, skin was mounted in side-by-side Franz diffusion cells (PermeGear, PA), with diameters of 0.9 cm and donor and receptor chamber volumes of 3.5 ml (Fig. 3.6). This well-established in vitro method is designed to closely mimic in vivo absorption [17]. The drug is applied to the skin surface facing the donor compartment either in solution or as a thin film, for example. The dermatomed porcine skin separates the two compartments. The receptor compartment is typically filled with a physiologically relevant buffer.

In these experiments, the donor comprised an aqueous solution of caffeine (20 mg ml^{-1}), a model drug, the passive skin permeation behaviour of which has been well characterised (through a normal intact barrier and across membranes compromised by adhesive tape-stripping [18] and by the use of arrays of microneedles [19, 20]). The receptor compartment contained the frequently used phosphate-buffered saline at pH 7.4 [21]. Both donor and receptor solutions were continuously, magnetically stirred. The cells were kept at a controlled temperature in an incubator at 37 °C, mimicking the in vivo skin surface temperature [1].

To track caffeine permeation, samples (1 ml) of the receptor solution were withdrawn hourly for 6 h and replaced with fresh buffer. The samples were filtered (Cronus syringe filter, Nylon 4 mm 0.45 μm, Labhut, Gloucester, UK) and then assayed for caffeine using high performance liquid chromatography (HPLC) (LC-2010 AHT, Shimadzu, MD)[22] with UV absorbance detection at 273 nm. The reverse phase HPLC column had an internal diameter of 4.6 mm and a length of 250 mm (HiQ sil C18HS, KYA Technologies, Tokyo, Japan). The mobile phase was 20 % acetonitrile, 0.1 % diethylamine in water and 79.9 % deionized water at pH 2.5. A flow rate of 1.0 ml min^{-1} was used and the retention time of caffeine was 5.7 min. Overall, the methods used followed closely those in the literature [18].

References

1. A. Williams, *Transdermal and Topical Drug Delivery from Theory to Clinical Practice* (Pharmaceutical Press, London, 2003)
2. J.E. Sader, J.W.M. Chon, P. Mulvaney, Calibration of rectangular atomic force microscope cantilevers. Rev. Sci. Instrum. **70**(10), 3967–3969 (1999)
3. J.E. Sader, Frequency response of cantilever beams immersed in viscous fluids with applications to the atomic force microscope. J. Appl. Phys. **84**(1), 64–76 (1998)
4. M.L. Oyen, R.F. Cook, Load-displacement behavior during sharp indentation of viscous-elastic-plastic materials. J. Mater. Res. **18**(1), 139–150 (2003)
5. T. Neumann, *Determining the Elastic Modulus of Biological Samples using Atomic Force Microscopy* (JPK Instruments AG, Berlin, 2009)
6. D.M. Ebenstein, L.A. Pruitt, Nanoindentation of biological materials. Nano Today **1**(3), 26–33 (2006)
7. M.L. Oyen, R.R. Cook, A practical guide for analysis of nanoindentation data. J. Mech. Behav. Biomed. Mater. **2**(4), 396–407 (2009)
8. J. D. Beard, S. N. Gordeev, Large flexibility of high aspect ratio carbon nanostructures fabricated by electron-beam-induced deposition. Nanotechnology **21**(47), (2010)
9. J.D. Beard, S.N. Gordeev, Fabrication and buckling dynamics of nanoneedle afm probes. Nanotechnology **22**(17), (2011)
10. S.L. Flegler, J.W. Heckman, K.L. Klomparens, *Scanning and Transmission Electron Microscopy: an Introduction* (Oxford University Press, New York, 1995)
11. K.L. Johnson, *Contact Mechanics* (Cambridge University Press, Cambridge, 1985)
12. A. L. Weisenhorn, S. Kasas, J. M. Solletti, M. Khorsandi, V. Gotzos, D. U. Romer, and G. P. Lorenzi. Deformation observed on soft surface with an AFM, volume 1855 of Scanning Probe Microscopies Ii. 1993
13. T.J. Ma, B. Bhushan, H. Murooka, I. Kobayashi, T. Osawa, A novel technique to measure the poisson's ratio and submicron lateral dimensional changes of ultrathin polymeric films. Rev. Sci. Instrum. **73**(4), 1813–1820 (2002)
14. D.C. Lin, D.I. Shreiber, E.K. Dimitriadis, F. Horkay, Spherical indentation of soft matter beyond the hertzian regime: numerical and experimental validation of hyperelastic models. Biomech. Model. Mechanobiol. **8**(5), 345–358 (2009)
15. L.A. Wegiel, L.J. Mauer, K.J. Edgar, L.S. Taylor, Crystallization of amorphous solid dispersions of resveratrol during preparation and storage-impact of different polymers. J. Pharm. Sci. **102**(1), 171–184 (2013)
16. N. Sekkat, R.H. Guy, Biological models to study skin permeation. *Pharmacokinetic Optimization in Drug Research* (2001), pp. 155–172
17. T.J. Franz, Percutaneous absorption—relevance of in vitro data. J. Invest. Dermatol. **64**(3), 190–195 (1975)
18. N. Sekkat, Y.N. Kalia, R.H. Guy, Porcine ear skin as a model for the assessment of transdermal drug delivery to premature neonates. Pharm. Res. **21**(8), 1390–1397 (2004)
19. R.F. Donnelly, T.R.R. Singh, M.J. Garland, K. Migalska, R. Majithiya, C.M. McCrudden, P.L. Kole, T.M.T. Mahmood, H.O. McCarthy, A.D. Woolfson, Hydrogel-forming microneedle arrays for enhanced transdermal drug delivery. Adv. Funct. Mater. **22**(23), 4879–4890 (2012)
20. M.J. Garland, K. Migalska, T.M. Tuan-Mahmood, T.R.R. Singh, R. Majithija, E. Caffarel-Salvador, C.M. McCrudden, H.O. McCarthy, D. Woolfson, R.F. Donnelly, Influence of skin model on in vitro performance of drug-loaded soluble microneedle arrays. Int. J. Pharm. **434**(1–2), 80–89 (2012)
21. S.W. Collier, N.M. Sheikh, A. Sakr, J.L. Lichtin, R.F. Stewart, R.L. Bronaugh, Maintenance of skin viability during in vitro percutaneous absorption-metabolism studies. Toxicol. Appl. Pharmacol. **99**(3), 522–533 (1989)
22. W.J. Lough, I.W. Wainer, *High Performance Liquid Chromatography : Fundamental Principles and Practice* (Blackie Academic & Professional, London, 1995)

Chapter 4
TEC Plasticization of Topical Polymeric Films

The degree of homogeneity and the mechanical properties of topical polymeric films for drug delivery are vital to their efficacy. The dispersal of plasticizer and the solubility of the drug within the films (the homogeneity of their distribution) affect their function; phase separations have been shown to trigger drug crystallization [1] and the efficacy (bioavailability) and stability of the drug are dependent on its crystallinity [2]. Inhomogeneities may also provide weak points in the films where fracture is more likely [3]. If the mechanical properties of the film are similar to those of skin, it will flex with the skin and remain in intimate contact for a prolonged period of time. The incorporation of plasticizers into polymer films increases their flexibility [4] and permits the mechanical properties to be matched to those of skin. In this way, film breakage is avoided and the area across which drug transport occurs remains constant.

AFM and Raman chemical mapping have previously been used to determine the drug/polymer miscibility in polymeric films [1, 5]. The occurrence of drug crystallization was detected by the appearance of aggregates and worm-like structures in AFM images. AFM nanoindentation has been used to determine the form of drug particles, e.g., crystalline, and to what degree, or amorphous [6–8]. Raman mapping has been used as a complementary method to AFM allowing direct detection, through peak shifts, of the degree of crystallinity [7].

This chapter describes the use of the AFM to acquire images of polymer films with and without a commonly used plasticizer, triethyl citrate (TEC), and betamethasone valerate (BMV), a model drug most commonly used for inflammatory skin conditions. Films of all compositions appear homogeneous with no obvious occurrence of phase separation or drug crystallinity. This supports the use of TEC as an effective plasticizer as it distributes evenly within the polymer network forming the film. The lack of crystallization in all films confirms that the presence of the plasticizer is not affecting the state of the drug.

AFM nanoindentation, to determine the material nature of the samples and their elastic moduli, was carried out. The samples showed viscous-elastic-plastic behaviour and elastic moduli were extracted using both the Hertzian model and an adapted

H. Garvie-Cook, *Novel (Trans)dermal Drug Delivery Strategies*,
Springer Theses, DOI 10.1007/978-3-319-28901-4_4

Oyen & Cook model, as described in Sect. 3.1.3.4. The calculated elastic moduli were found to decrease with increasing plasticizer content.

Raman spectroscopy of the constituents of the polymeric films (polymers, the plasticizer and the drug substance) revealed characteristic spectral peaks by which they could be identified. These peaks were used to determine the distribution of these components from the measured Raman chemical maps of films containing plasticizer and the drug. This information was on a larger scale (with a lower resolution) than that obtained using AFM but was complementary, showing the homogeneity of the plasticizer and drug distribution within the films.

4.1 AFM Imaging

Images of polymeric FFS, cast on cleaned glass slides, were acquired in AFM tapping mode. Both Eudragit and Klucel films, incorporating 0, 20 and 40 % TEC, were prepared and the images of these films are shown in Fig. 4.1.

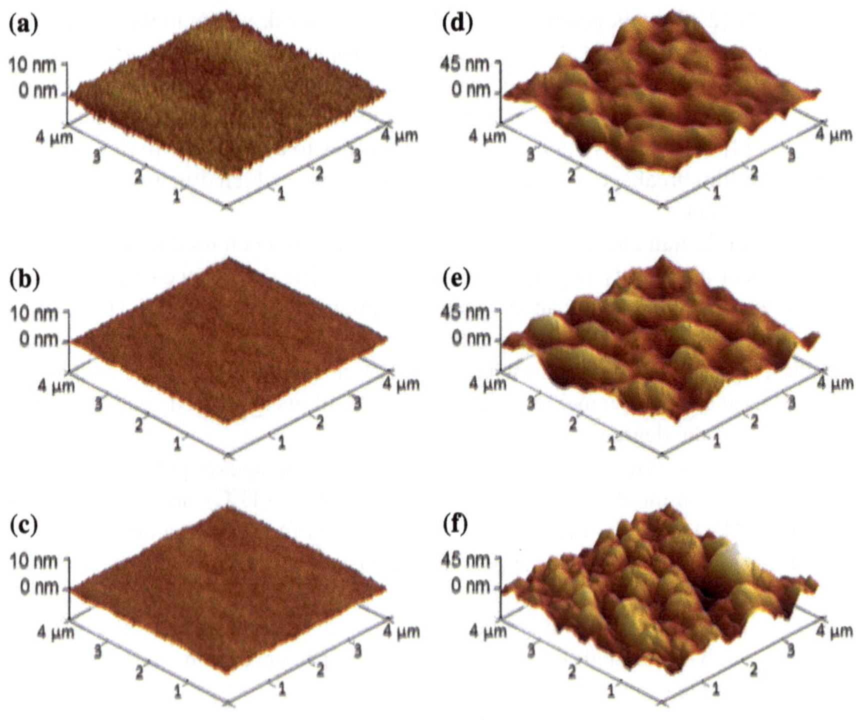

Fig. 4.1 3D AFM images of polymeric films with 0, 20 and 40 % TEC plasticizer, cast onto glass microscope slides, at scan sizes of $4 \times 4\ \mu m^2$: **a–c** Eudragit films with 0, 20 and 40 % TEC, respectively, and **d–f** Klucel films with 0, 20 and 40 % TEC, respectively

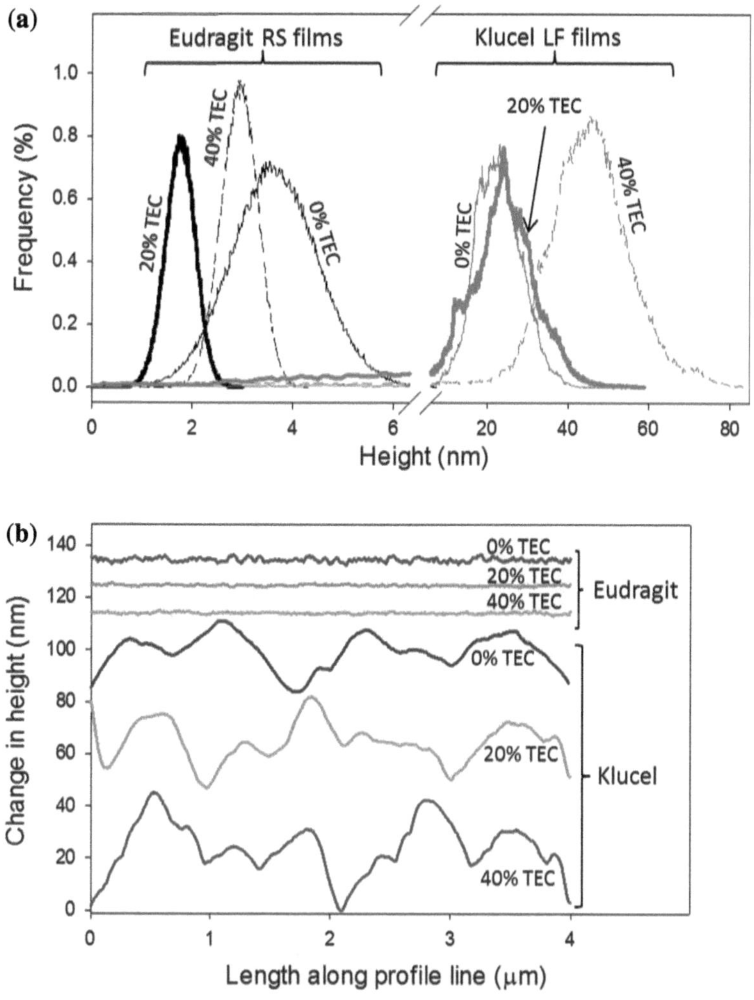

Fig. 4.2 **a** Histograms of AFM image feature height frequency, and **b** the change in height along one line of AFM images of films incorporating 0, 20 and 40 % TEC plasticizer. Height profiles in (**b**) are offset to facilitate comparison

Histograms of the heights of features in AFM images (Fig. 4.2a) represent the range of features observed. These histograms were produced by determining the number of features over certain threshold heights. Height profiles along one line of the image are also shown (Fig. 4.2b) to further support the information in the histograms.

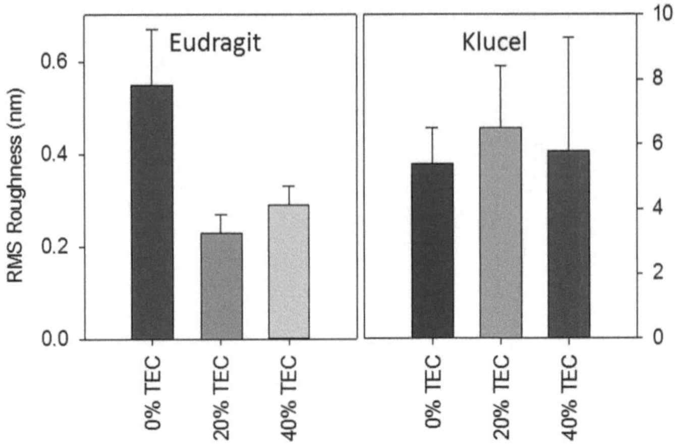

Fig. 4.3 RMS roughness (+ standard deviation) of Eudragit and Klucel films, with 0, 20 and 40 % TEC, determined from three AFM tapping mode images of each film

Root mean square (RMS) average roughness provides another description of the features observed in AFM images. It is the average of height deviations from a mean image data plane.

Eudragit films without TEC (Fig. 4.1a) show small structural features, with heights just below 4 nm (Fig. 4.2). The heights of these features decrease with increasing TEC content and the films were smoother when 20 and 40 % TEC (Fig. 4.1b, c) were incorporated. The RMS roughness (R_q) of three films of each composition, cast in the same way, decreased from 0.55 ± 0.12 nm, in films without TEC, to 0.23 ± 0.04 and 0.29 ± 0.04 nm in films with 20 and 40 % TEC, respectively. The changes in RMS roughness with increasing plasticizer content are shown in Fig. 4.3.

Klucel films without TEC show larger structural features than those observed in the equivalent Eudragit film, with heights of approximately 20 nm (Fig. 4.2). The root mean squared roughness of this film is 5.4 ± 1.1 nm. As the content of TEC increases, the heights of the structural features increase to approximately 45 nm in films incorporating 40 % TEC (Fig. 4.2a), but the RMS roughness did not change: 6.5 ± 1.9 and 5.8 ± 3.5 nm for films incorporating 20 and 40 % TEC, respectively (Fig. 4.3).

Polymeric films form from the FFS as the solvent evaporates from the solution. Ethanolic solutions of both Eudragit and Klucel are clear and are therefore believed to be polymeric solutions, rather than dispersions. In solution, polymer chains are highly mobile but come into close contact as the solvent evaporates, eventually forming a film as shown below. The rate of film formation and the structure of the formed film depend on the rate of solvent evaporation and the presence (or not) of a plasticizer.

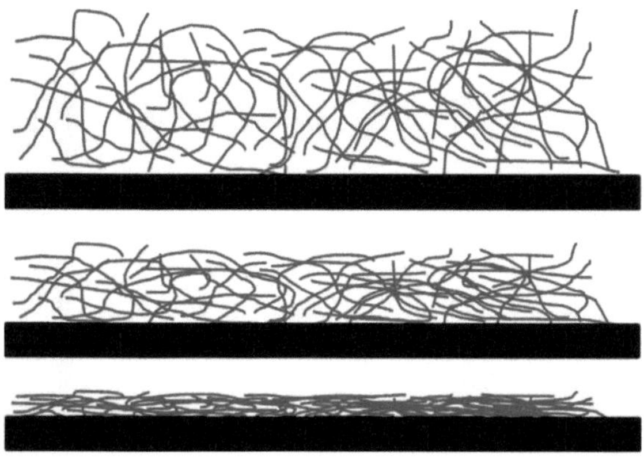

As the solvent evaporates, the mobile polymer chains move closer together until they form rigid polymer-polymer contacts. An effective plasticizer increases the flexibility of the film by decreasing the number of polymer-polymer contacts and therefore the rigidity of the formed structure [3, 9–11]. Ideally, the drug is either dissolved in the polymer or in an amorphous state within the polymeric network [12, 13].

There are no signs of phase separation or drug crystallization in the AFM images of any film, indicating that, when present, TEC and BMV were distributed evenly throughout the film. TEC has therefore effectively plasticized the formed films. The observed homogeneity of the films is an important quality as inhomogeneous polymer coatings can cause more rapid drug release from parts of the film [10]; equally, the formation of drug crystals decreases the eventual bioavailability of the drug [1].

The roughness of Eudragit films (with and without drug) decreases with increasing plasticizer content. The presence of plasticizer in the polymer network provides “gap-fillers”, which result in a smoother film [5]. The same decrease in roughness with plasticizer content was not observed for Klucel films which did not contain the drug substance. This may be due to the larger molecular weight of the Klucel (95,000 g mol^{-1}), relative to the plasticizer (276 g mol^{-1}), and compared to Eudragit (32,000 g mol^{-1}), resulting in a lower change in roughness.

4.2 AFM Nanoindentation

4.2.1 Deformation Behaviour

Nanoindentation of films of different composition was carried out using AFM probe tips. The raw data from the indentation procedure (deflection in volts as a function of displacement of the AFM cantilever from its original vertical position) was converted to the deformation of the sample according to the applied load using the procedure

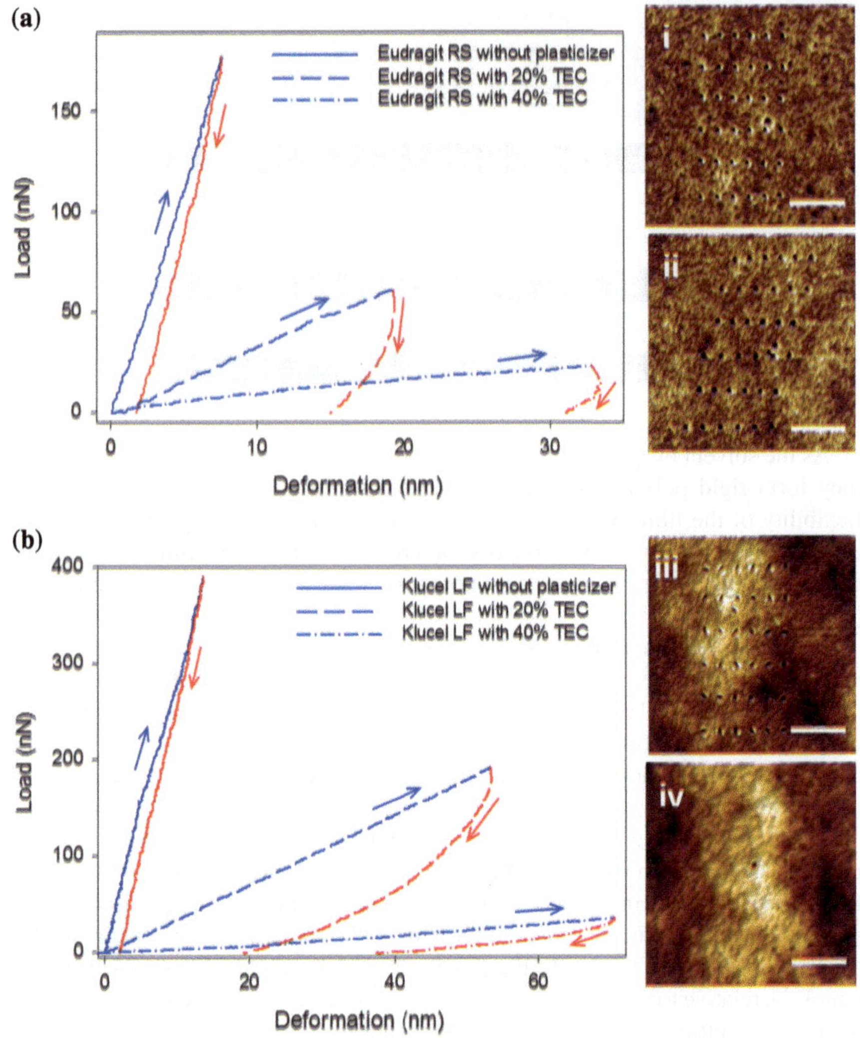

Fig. 4.4 Load as a function of deformation during indentation of films with and without plasticizer (TEC): **a** Eudragit with 0, 20 and 40 % TEC, and **b** Klucel with 0, 20 and 40 % TEC. *Blue* and *red lines* are data collected as the load was increased and decreased, respectively. 2D AFM images of (*i*–*ii*) Eudragit with 0 and 20 % TEC, and (*iii*–*iv*) Klucel with 0 and 20 % TEC were taken after indentation. Scale bar represents 0.5 μm. Indents are less obvious in (*iv*) due to the more viscous nature, and the greater roughness of the sample, as compared to (*ii*)

described in Sect. 3.1.3. Examples of the resulting data from samples incorporating 0, 20 and 40 % TEC (Fig. 4.4) clearly reveal the impact of plasticizer incorporation. The greatest load is required to deform (and the steepest loading curve is obtained for) Eudragit films without TEC (Fig. 4.4a). As TEC content increases, the load required to deform the sample by a given amount decreases, showing the softer nature of films. Similar behaviour, with increasing TEC content, was observed for Klucel films (Fig. 4.4b).

The variation of deformation with load on the polymeric films reveals the indentation behaviour of the samples. This allows a model to be selected that accounts for the deformation behaviour of the samples and allows determination of their elastic moduli [14]. Large hysteresis, and a residual deformation when the load has been removed, were observed (Fig. 4.4), implying that some plastic deformation occurred during indentation. This residual deformation can be observed in AFM images of the films taken after indentation (Fig. 4.4i–iv).

4.2.1.1 Viscosity

Viscous behaviour can be assessed using the unloading data obtained during indentation. For Eudragit films with 20 and 40 % TEC, the hysteresis between the loading and unloading data remains large until the load is entirely lifted and a residual deformation is left (Fig. 4.4a). In Klucel films with 20 and 40 % TEC, this hysteresis is lower and the depth of the residual deformation left after the load is lifted is smaller (Fig. 4.4b). This difference shows the greater reversibility (i.e., the more elastic nature) of Klucel films containing TEC than the corresponding Eudragit films.

Viscous deformation was investigated by varying approach rate and surface delay. The approach rate is the speed at which the AFM probe is brought down towards and pushed into the sample. The probe is then retracted from the sample at the same speed. Approach rates of 4 and 16 nm s^{-1} were used for the indentation of all films investigated. Example indentation loops acquired at these different approach rates are shown in Fig. 4.5.

Less deformation, during loading and unloading, was observed for indents carried out at approach rate of 16 nm s^{-1} than those using 4 nm s^{-1}. Using Fig. 4.5 as an illustrative example, at a load of 30 nN, the deformation during loading of the Eudragit film without plasticizer was lower (1.6 nm compared to 2.2 nm) at the higher approach rate. This is also reflected in the behaviour of Klucel films with 40 % TEC; the deformations during loading were 76 and 121 nm for approach rates of 16 and 4 nm s^{-1}, respectively. When approaching at a higher rate, the sample provides a greater resistance to deformation as there is less time for viscous relaxation to occur [15]. The deformation at a given load is therefore lower.

There was less hysteresis between the approach and retract curves when a higher approach rate was used. In Fig. 4.5, the difference in deformation between the approach and retract curves at a load of 30 nN was 2 and 5 nm at approach rates of 16 and 4 nm s^{-1}, respectively for Eudragit films without plasticizer. For Klucel films with 40 % TEC, the hysteresis between the approach and retract curves at a load of 30 nN was 44 and 51 nm at approach rates of 16 and 4 nm s^{-1}, respectively. Again, this difference is due to the time allowed for more viscous relaxation to occur at a slower approach rate.

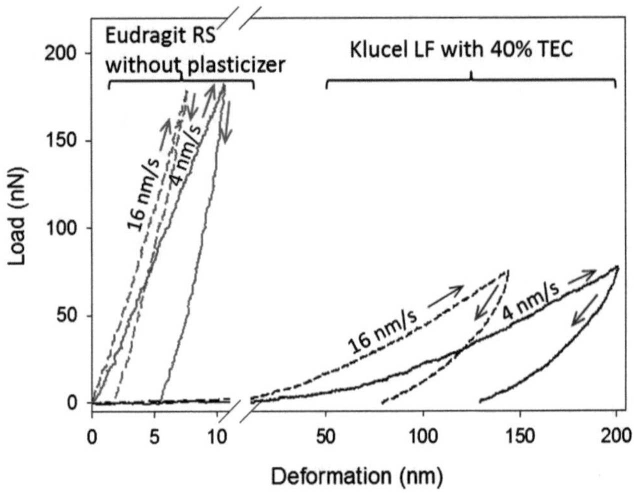

Fig. 4.5 Load as a function of deformation during indentation of a Eudragit film without plasticizer, and a Klucel film incorporating 40 % TEC, at approach rates of 4 nm s^{-1} (*solid lines*) and 16 nm s^{-1} (*dashed lines*)

Another indication of viscous deformation is the appearance of a "nose" in the initial unloading data of the samples. Even during unloading, the deformation of the sample may still increase (the "nose" in the data) if the sample shows viscous behaviour. In this case, the rate that the sample creeps under the applied load (due to its viscosity) is greater than the unloading rate. An example of this initial increase in deformation during unloading can be seen in Fig. 4.4a: the viscous nature of Eudragit films with 40 % TEC is sufficient that the probe continues to sink into the sample even after the load has begun to be lifted. The gradient of the initial unloading portion can therefore be used to assess the viscosity of the sample.

The gradient of the "nose" was calculated for the data shown in Fig. 4.4. The values for the initial unloading portions (the first 25 % of the data) for Eudragit films containing 0, 20 and 40 % TEC were 36, 66 and –15 nN/nm, respectively. As the plasticizer content of the films increases, the probe sinks into the sample at a greater rate during unloading. This causes the gradient to increase and then to become negative. Eudragit films with 40 % TEC therefore show a greater viscous nature than samples without plasticizer. The gradients of the initial unloading portions of Klucel films were also calculated. The values were 38, 15 and 2 nN/nm for films incorporating 0, 20 and 40 % TEC, respectively. In this case, the gradient becomes shallower, i.e., less unloading is required for a given decrease in deformation, as expected from the softer nature of samples containing TEC, and an obvious "nose" is not observed. Eudragit films therefore show more viscous deformation than Klucel films when TEC is incorporated. The molecular weight of Eudragit is lower than that of Klucel (32,000 compared to 95,000 g mol^{-1}) and the degree of its polymer chain entangle-

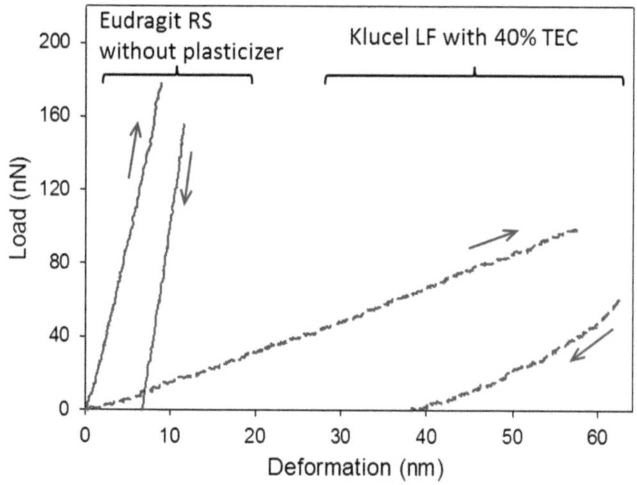

Fig. 4.6 Load as a function of deformation during indentation of a Eudragit film without plasticizer (*solid lines*), and a Klucel film incorporating 40 % TEC (*dashed lines*). A delay of 10 s at maximum load and an approach rate of 16 nm s^{-1} were used in these measurements

ment is less. This results in greater chain mobility and the more viscous deformation observed in Eudragit films is emphasized when TEC is incorporated [13].

Delays of 0 and 10 s were used to assess the creep of the probe during indentation. During this delay, the cantilever is held at its maximum vertical displacement, at which point the probe tip imposes the greatest load on the sample. In purely elastic-plastic samples, deformation would not change during this delay. When a surface delay of 10 s was used during the indentation of Eudragit films without plasticizer and Klucel films with 40 % TEC, both of which show viscous behaviour, an increase in the deformation of the sample at the maximum load is observed (Fig. 4.6). During the delay, the probe sinks further into the sample, even though the cantilever is being held at a fixed position. The deflection of the cantilever, and therefore the measured load, decreases during the delay as the deformation caused by the samples' viscosity increases. In the case of Eudragit films without plasticizer, the load decreases to 87 % of its original value before the hold period. The load decreases more, to 61 % of its original value, for Klucel films with 40 % TEC. This is consistent with the more viscous nature of the sample containing TEC plasticizer. An increase in the viscous behaviour of polymeric films with increasing plasticizer content has previously been observed in hydroxypropylmethylcellulose films plasticized with glycerol and polyethylene glycol [3]. This was attributed to the reduction in rigid polymer-polymer contacts when plasticizer is incorporated.

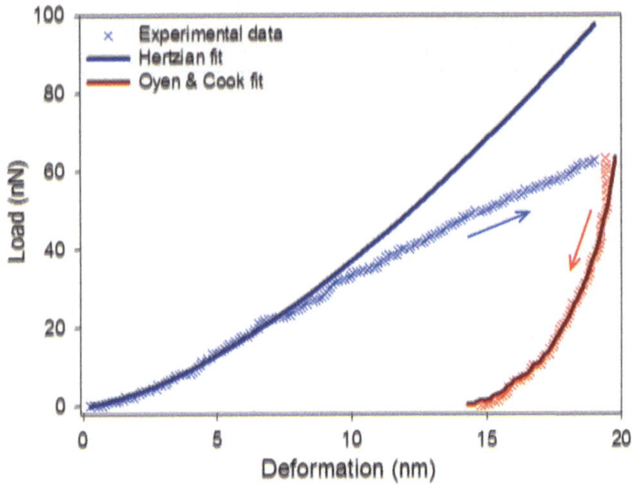

Fig. 4.7 Fits of the Hertzian and Oyen & Cook models (*upper* and *lower solid lines*, respectively) to the experimentally measured load as a function of deformation (*crosses*) during indentation of a Eudragit film incorporating 20 % TEC

4.2.2 Extraction of Elastic Modulus

4.2.2.1 Comparison of Hertzian and Oyen and Cook Models

The elastic moduli of polymeric films of different compositions were determined using the methods described in Sect. 3.1.3.4. Both Hertzian and Oyen & Cook models were fitted to experimental indentation data, and examples are shown in Fig. 4.7.

The Hertzian model describes the behaviour of purely elastic media. As shown in Sect. 4.2.1, the polymeric films under investigation are not purely elastic. However, for very small deformations, the Hertzian model is often considered to be a reasonable approximation for samples which are not purely elastic.

Loading data within the initial 5–20 % of experimental points was fitted to extract the elastic modulus using the Hertzian model. The deformation here was assumed to be essentially elastic. It is noted that the range chosen excludes the initial loading data when the indentation process is most prone to noise [16]. This also resulted in the most consistent results with sufficient data points for all samples to provide a reasonable fit. The Hertzian model explains well the first 5–20 % of data but then diverges predicting a lower deformation at a given load than is seen experimentally. This is expected as only the elastic deformation of the sample is modelled while, at greater loads, plastic deformation and viscous deformation become more apparent. The experimental deformation of the sample at a given load is therefore higher than that predicted by the purely elastic Hertzian model.

The Oyen & Cook model, which takes into account the viscous-elastic-plastic behaviour of the sample, was fitted to the central 75 % of experimental data. Elastic

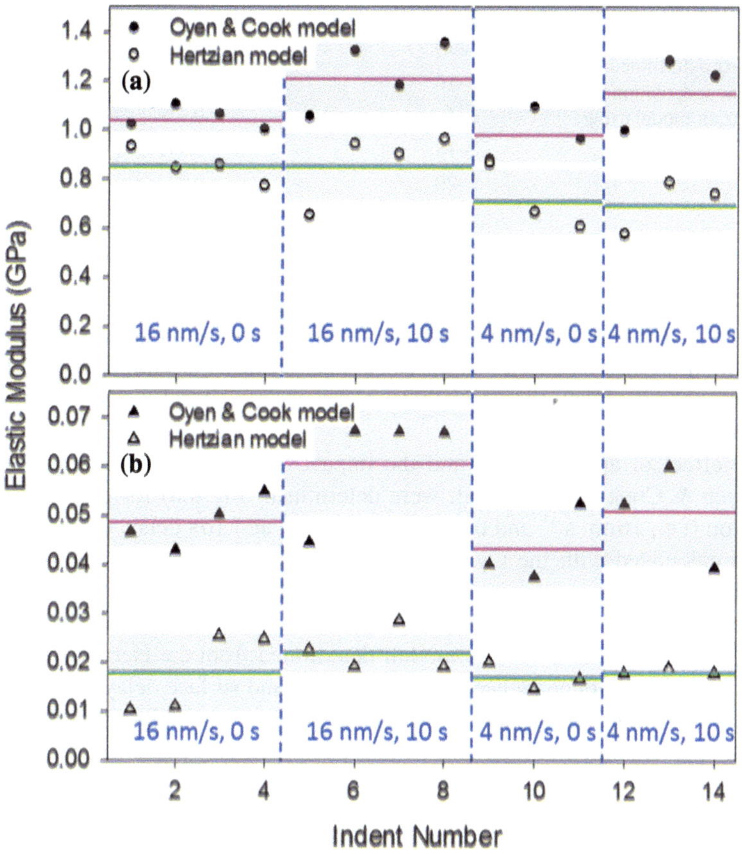

Fig. 4.8 Examples of Oyen & Cook and Hertzian model-calculated elastic moduli for (**a**) Eudragit without plasticiser, and (**b**) Klucel incorporating 40 % TEC, determined under four different indentation conditions: approach rate of 16 nm s^{-1} and 0 s delay, 16 nm s^{-1} and 10 s, 4 nm s^{-1} and 0 s, 4 nm s^{-1} and 10 s. Averages of Oyen & Cook and Hertzian moduli for each indentation condition are shown as *pink* and *green lines*, respectively. Areas shaded in grey represent the standard deviations of these averages

moduli values calculated using the Oyen & Cook model were consistently higher than those obtained from the Hertzian model (the comparison between the results for two different films is shown in Fig. 4.8). It follows that the data range fitted by the Hertzian model included some non-linear plastic and viscous deformation and the model therefore predicted a greater elastic deformation than was actually occurring (and hence resulted in a lower elastic modulus). This was verified by varying the data range over which the elastic modulus was determined (an example of this is shown in Table 4.1). The narrower the range of data points fitted (i.e., those closest to the start of the indent), the higher the derived elastic modulus. The deformation at these initial points is more likely to be purely elastic and the elastic modulus extracted is higher, and in agreement with the Oyen & Cook prediction. As the fitted data range

Table 4.1 Elastic modulus of a Eudragit film without plasticizer determined from fits of one indentation loop to the Hertzian model using different data ranges

Fitted range (% of loading data)	Hertzian elastic modulus (GPa)
1–16	1.04
3–18	0.972
5–18	0.941
5–20	0.927
5–30	0.887
5–40	0.823

increases and includes more points from the later stages of the indent, the Hertzian elastic modulus decreases. Here, the viscous and plastic contributions to deformation are greater and it is likely, therefore, that the Hertzian model underestimates the elastic modulus of the sample.

The effect of approach rate and the impact of a delay on the fitted Hertzian and Oyen & Cook elastic moduli were determined (Fig. 4.8) for each indentation condition (i.e., 16 nm s^{-1} and 0 s delay, 16 nm s^{-1} and 10 s delay, etc.). The elastic moduli calculated with the Oyen & Cook model were not significantly different over all the indentation conditions used. This model successfully accounts therefore for the viscous and elastic deformations, observed in Fig. 4.4, that occur during unloading. The values of elastic moduli determined from the Hertzian model also showed no obvious dependence on approach rate and surface delay suggesting that the contributions of viscous and plastic deformation are insignificant in the range of experimental conditions examined.

The variation of the Hertzian and Oyen & Cook model-calculated elastic moduli as a function of the maximum (peak) load used during indentation is shown in Fig. 4.9

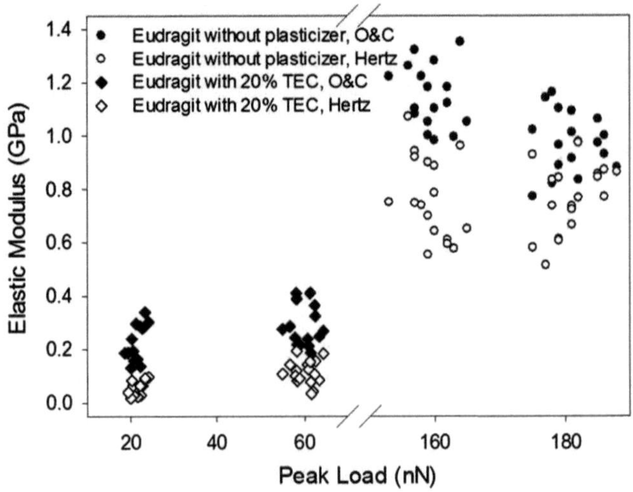

Fig. 4.9 Examples of Oyen & Cook and Hertzian elastic moduli for Eudragit without plasticizer and Eudragit with 20 % TEC as a function of the peak load imposed during indentation

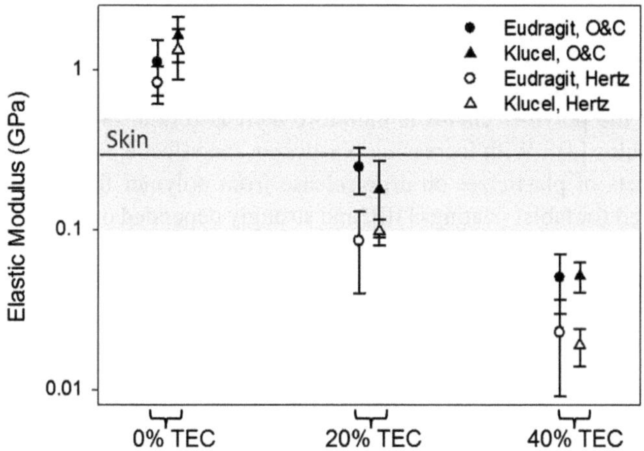

Fig. 4.10 Oyen & Cook and Hertzian model-calculated elastic moduli of Eudragit and Klucel films without and with 20 and 40 % plasticizer. The data shown are the means and standard deviations determined from three different samples of each film and a minimum of eight indents per sample. The elastic modulus of skin [17], determined under similar indentation conditions, is shown as a *horizontal line* for comparison, with the standard deviation shown in *grey*

for Eudragit films with and without 20 % TEC. The Hertzian elastic moduli did not vary with peak load for either film, behaviour characteristic of an ideal elastic material [7]. Elastic modulus calculated using the Oyen & Cook model appeared to decrease somewhat at higher loads, where the viscous deformation of the sample will become more prevalent. This suggests that the model may not have completely accounted for the viscous behaviour of the films.

Finally, all elastic moduli calculated for all films investigated were unchanged when the drug substance (1.2 % w/w) was incorporated (data not shown).

4.2.2.2 Variation with Plasticizer Content

Elastic moduli of Eudragit and Klucel films, calculated using Hertzian and Oyen & Cook models, as a function of TEC content, decreased with increasing plasticizer content, as expected (Fig. 4.10) [4]. Although the absolute, calculated values of the two models differed, the large decrease in modulus with the incorporation of plasticizer is patently clear.

Similar behaviour has been observed in other films incorporating plasticizer: a decrease in elastic modulus in hydroxypropylcellulose hot-melt extruded films when plasticized using polyethylene glycol [18], an increase in flexibility and elasticity when polyvinyl alcohol films incorporated glycerine as a plasticizer [19], and a reduction in the tensile strength of Eudragit® L 100-55 films plasticized with TEC [20]. The elastic modulus of human skin, approximately 0.3 GPa [17], is lower than the moduli of unplasticized Eudragit and Klucel films. A plasticizer content of 20 %

TEC must be introduced into the films to bring their elastic moduli down to that of skin and to ensure intimate and prolonged contact between the skin and the topically applied film. Plasticizer within the film decreases polymer-polymer contact, and the mobility of the polymer chains is therefore increased (and the films have a lower elastic modulus [4]). With increasing plasticizer, the effect is magnified.

The effects of plasticizer on drug release from polymer films have previously been reported for tablet coatings [10], and strongly depended on the plasticizer used (hydrophilic or lipophilic). Drug release from films of the same composition as those studied here increased with with increasing TEC content [21], presumably due to enhanced difusivity within the polymer network [13].

4.3 Raman Micro-Spectroscopy

4.3.1 Constituent Spectra

Raman spectra of the constituents of polymeric FFS were initially acquired (Fig. 4.11) to determine characteristic peaks by which they could be identified in spectroscopic maps of the formed films. Characteristic peaks for BMV and TEC were identified at 1659 and 1734 cm^{-1}, respectively. The peak for BMV was distinct from any polymer or plasticizer peak and could therefore be used to map the drug within films of both polymers. The 1742 cm^{-1} peak for TEC could be clearly distinguished in the Klucel

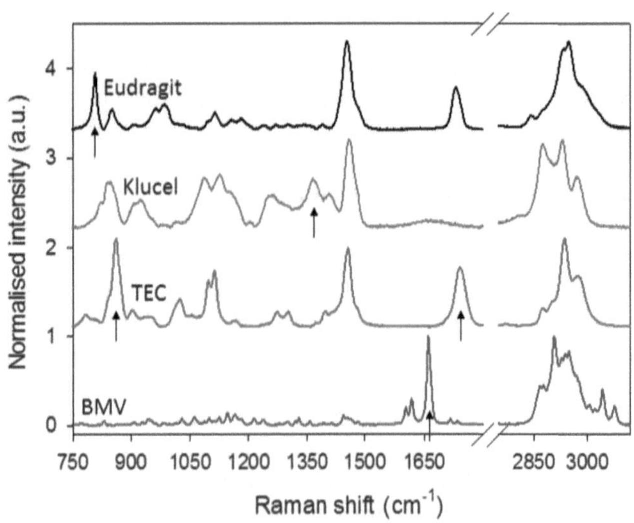

Fig. 4.11 Raman spectra of the constituents of polymeric films: Eudragit, Klucel, TEC and BMV. Spectra are normalised according to the maximum intensity within the two ranges shown. *Arrows* indicate the positions of characteristic peaks for each of the constituents

films but overlapped with a Raman signal from the polymer in Eudragit films. In these films, therefore, the peak intensity at $857\,cm^{-1}$ was used to map the distribution of TEC. Peaks which were characteristic of the polymers Eudragit and Klucel were identified at 813 and $1358\,cm^{-1}$, respectively.

Raman chemical mapping was performed at a high spatial resolution to provide comparison with AFM images. The Raman signal obtained during mapping originated from an area of approximately $1 \times 1\,\mu m^2$. Maps were acquired of Eudragit and Klucel films, incorporating 20 % TEC and 1.2 % BMV, respectively. Spectra from the maps were processed by subtracting the background. The software used assumed the same background, as defined in one spectrum, for every pixel of the map.

The position of the characteristic peaks of TEC shifted when the plasticizer was incorporated into Eudragit and Klucel films, as opposed to in its powder form. Over the mapped area of Eudragit and Klucel films with 20 % TEC and 1.2 % BMV, the TEC peak positions were $858.3 \pm 0.1\,cm^{-1}$ and $1729.5 \pm 0.2\,cm^{-1}$, respectively. In both cases, therefore, the position of the peak decreased, suggesting that the environment in which the scattering TEC molecules were contained was altered. As no inhomogeneities were observed in either AFM images or Raman concentration maps (see Sect. 4.3.2), it is likely therefore that TEC was dispersed within the network provided by the film-forming polymers.

A similar effect was observed when the drug was incorporated into polymeric films; the characteristic peak of BMV shifted to 1669.0 and $1666.9\,cm^{-1}$ in Eudragit and Klucel films (without TEC), respectively. This is further discussed in Sect. 4.3.3.

4.3.2 Mapping Concentration

The focus of the Raman microscope objective, and therefore its ability to collect the scattered radiation, is affected by changes in the height of the film across the mapped area. If the sample becomes out of focus, a lower intensity and background will be collected. Maps of the intensities of characteristic peaks of the polymer, plasticizer and drug within Eudragit films with 20 % TEC and 1.2 % BMV were similar, suggesting that changes in the topography of the film, rather than changes in the film's chemical distribution, explained the distribution observed.

To account for changes in sample height, spectra acquired along a line of each map were analysed. The intensities of characteristic peaks of polymer, plasticizer and drug were determined for each of these spectra then normalised according to each component's maximum intensity in the line of spectra analysed (Fig. 4.12a). The intensities of all characteristic peaks decreased to a minimum at a distance of approximately 9 μm along the line and then increased progressively to a maximum value observed at approximately 26 μm.

Prior to the analysis of peak intensities, background was subtracted from the individual spectra which formed the map. The Wire software, which was initially used to subtract background, assumed the same background spectrum over the mapped area. The variation in background intensities due to sample focus were therefore not

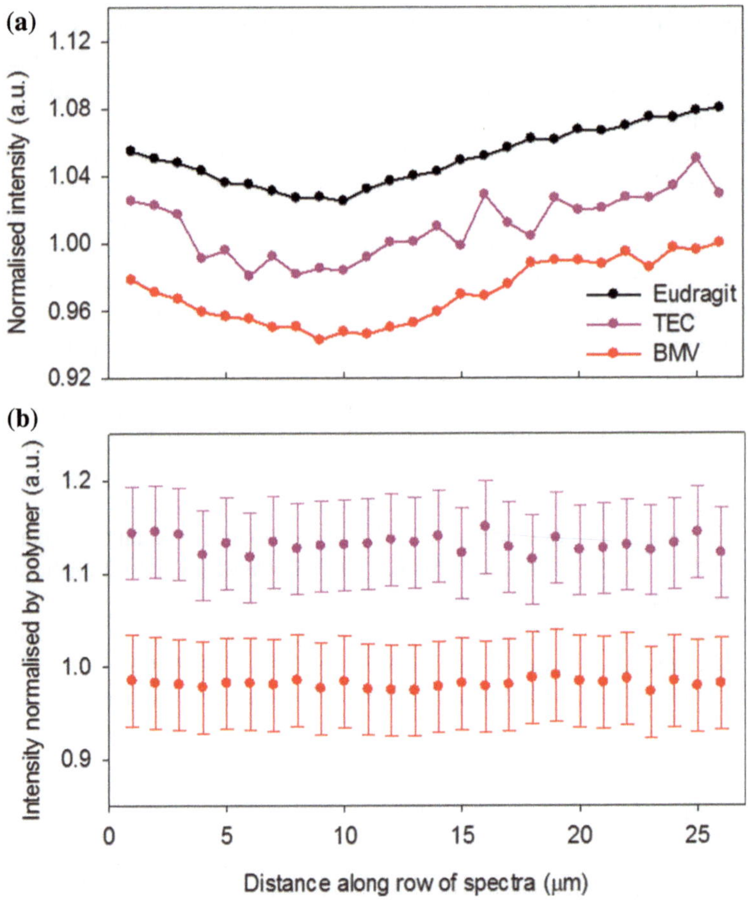

Fig. 4.12 **a** Raman peak intensities of Eudragit, TEC and BMV along a row of 26 spectra of a Eudragit film incorporating 20 % TEC and 1.2 % BMV. Intensities were normalised according to their maximum along this row. **b** Peak intensities of TEC and BMV were then divided by the intensities of the polymer peak. The maximum intensities of these divided signals were then defined as 1. Errors were calculated using the standard deviation of TEC and BMV peak intensities after repeating manual background subtraction five times on one spectrum. Intensities are offset to facilitate comparison

accounted for. After the deduction of background, spectra from areas of the film which were more in focus had higher intensities, thought to explain the variation in intensity observed in Fig. 4.12a.

The polymer was assumed to be distributed evenly throughout the film and the intensities of the spectra along the line investigated were therefore divided by the polymer signal, to account for the change in background (Fig. 4.12b). To estimate an error on the intensity of characteristic peaks, the background signal of a spectrum of each film was defined, without using the automatic function in the Wire software, and

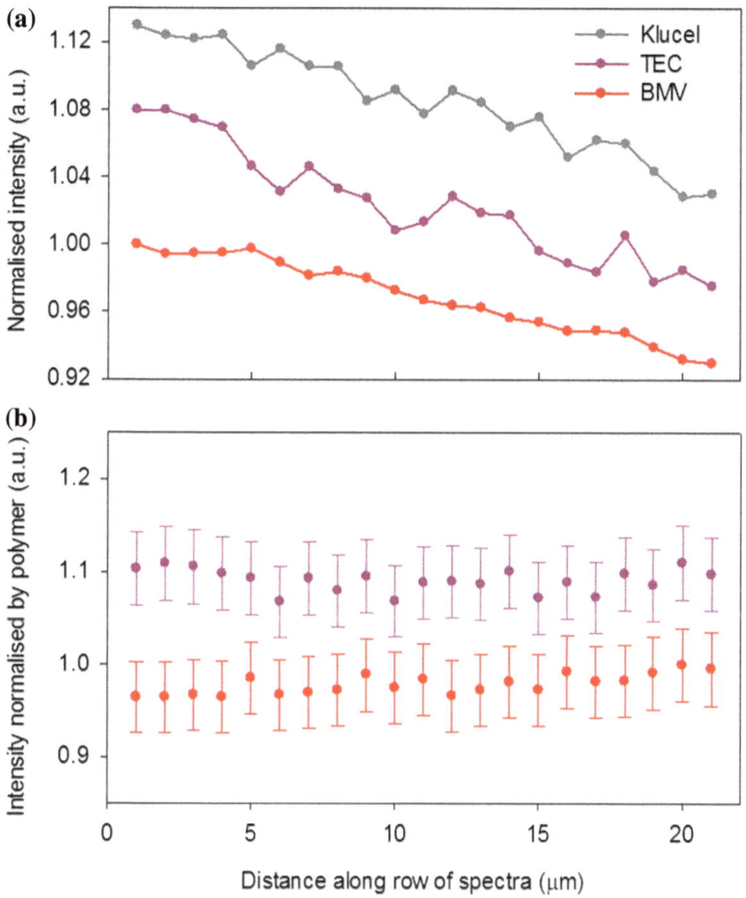

Fig. 4.13 **a** Raman peak intensities of Klucel, TEC and BMV along a row of 21 spectra of a Klucel film incorporating 20 % TEC and 1.2 % BMV. **b** Peak intensities of TEC and BMV were then divided by the intensities of the polymer peak. The maximum intensities of these divided signals were then defined as 1. Errors were calculated using the standard deviation of TEC and BMV peak intensities after repeating background subtraction five times on one spectrum. Intensities are offset to facilitate comparison

subtracted. Intensities of characteristic peaks were then deteremined and this process was repeated five times. The standard deviations of the determined intensities were used to calculate percentage errors. The concentrations of plasticizer and drug over the row of spectra from Eudragit with 20 % TEC and 1.2 % BMV were effectively constant (Fig. 4.12b) and TEC and BMV were therefore distributed evenly.

The same analysis was applied to maps of Klucel with 20 % TEC and 1.2 % BMV. In this case, the intensity of all characteristic peaks decreased in the same manner with increasing distance along the line of spectra, suggesting that the sample was became progressively out of focus (Fig. 4.13a). The polymer was again assumed to be evenly distributed and the intensities of plasticizer and drug peaks were normalised according to its signal. The error was calculated in the same way as described above. The intensities of plasticizer and drug along the row of spectra analysed were, again, effectively constant (Fig. 4.13b). In both Eudragit and Klucel films incorporating TEC and BMV, plasticizer and drug were therefore distributed evenly, supporting the homogeneity observed in AFM images (Fig. 4.1).

4.3.3 Mapping the Physical State of the Drug

BMV was assessed spectroscopically in three physical forms and its characteristic peak (at spectral shifts of approximately 1650–1670 cm^{-1}) was determined. The three forms were crystalline (as provided by the supplier), as an amorphous film, formed by depositing a drop of an ethanolic solution onto a clean glass slide [22], and in solution (0.4 % w/w in ethanol), and the spectra acquired are in Fig. 4.14. The peak shifted from 1659 and 1663 cm^{-1} for crystalline and amorphous solid forms, respectively, to 1666 cm^{-1} for the dissolved compound.

The physical state of BMV across the films was then determined by mapping the frequency of its characteristic peak. In Eudragit films, the peak was observed at

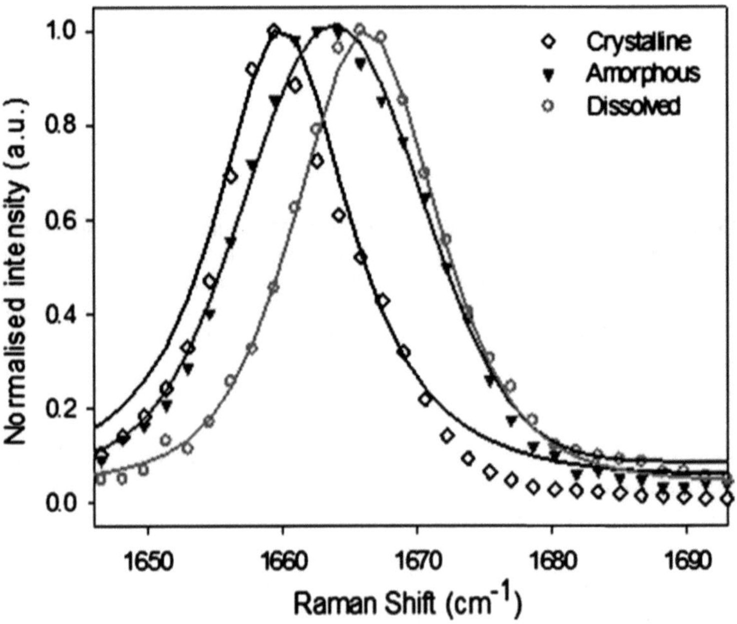

Fig. 4.14 Raman spectra of crystalline, amorphous and dissolved BMV. Spectra have been normalised to a maximum value of 1 within the range shown. Gaussian functions (*solid lines*) were fitted to spectral data points and used to determine the peak position

1667.8–1667.9 cm^{-1}; that is, a range much smaller than the difference, for example, between amorphous and dissolved drug, suggesting that BMV was dissolved and equally soluble throughout the mapped area of the film. The corresponding range of the peak position in a Klucel film incorporating 20 % TEC and 1.2 % BMV was 1665.9–1666.0 cm^{-1}, implying that BMV was dissolved but consistently less soluble across the mapped area. The positions of the BMV peaks were lower than those observed in Eudragit and Klucel films without TEC, suggesting that the incorporation of TEC has slightly reduced the solubility of the drug within the polymer film.

4.4 Film Formation on Skin

Section 4.2 describes the deformation behaviour and elastic moduli of topical films which were cast on glass slides and dried for 24 h at 30 °C. To determine whether the observed properties would vary when in situ on the surface of skin, AFM measurements of Eudragit films incorporating 0, 20 and 40 % TEC, prepared in three different ways, were carried out. The effects of the conditions of deposition and the substrate, i.e., glass or skin, were determined.

To examine the impact of changing the preparation method, films were cast onto either glass slides or excised porcine skin and dried for 6 hours at either room temperature or 30 °C. Example AFM indentation loops for these samples are in Fig. 4.15. The probe used for indentation had a smaller radius than those used before: 7 nm as compared to 21–80 nm. When a smaller probe is used, more plastic deformation occurs at a smaller load, making direct comparison of Fig. 4.15 with Fig. 4.4 less than straightforward. In general, however, the shapes of all of the load-deformation plots in these experiments were similar. For example, indentation of a Eudragit film incorporating 20 % TEC revealed similar viscous and plastic characteristics regardless of the preparation method used. Deposition and substrate parameters do not, therefore, appear to change the deformation behaviour of the films.

Variation in the films' mechanical properties, when prepared differently, can be quantified via their elastic moduli. As a sharper probe was used, more viscous and plastic deformation was observed, resulting in large hysteresis and residual deformation in all indentations. Elastic moduli were calculated by fitting the Hertzian model to small deformations; i.e., to the first 5–20 % of data acquired during loading (Fig. 4.16). Increasing the plasticizer content of the films resulted in a decrease in elastic modulus, as previously observed (Fig. 4.10). The preparation method had the most significant effect on the elastic moduli of Eudragit films with 40 % TEC, with those prepared at room temperature (on both glass and skin) having lower values than those prepared at 30 °C.

Elastic moduli of Eudragit films without plasticizer, and with 20 % TEC, prepared on glass slides at both 30 °C and room temperature, were higher than those which had been dried for 24 h at 30 °C. This difference is likely due to the smaller probe radius used for indentation. Nonetheless, the elastic moduli of the films were still observed to significantly decrease with increasing TEC.

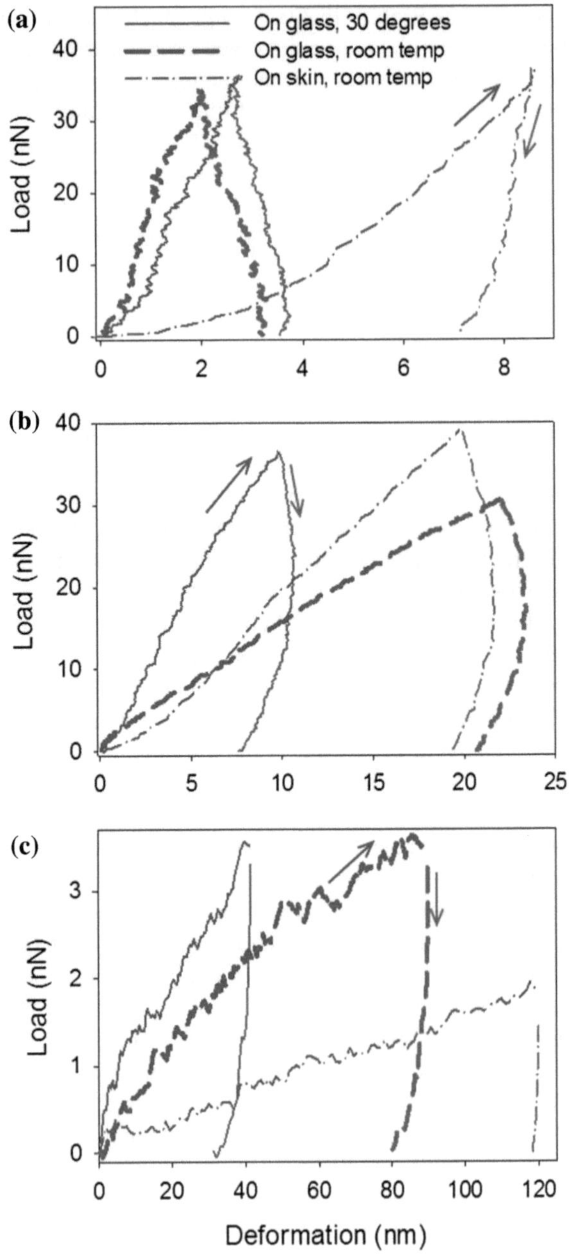

Fig. 4.15 Examples of load as a function of deformation during indentation of Eudragit films **a** without plasticizer, **b** with 20 % TEC, and **c** with 40 % TEC. Films were dried for 6 h on glass at 30 °C (*solid lines*), on glass at room temperature (*thick dashed lines*), and on skin at room temperature (*dash-dot-dashed lines*). *Blue* and *red lines* represent the data collected as the load was increased and decreased, respectively, as indicated by the arrows (color online)

Under the same load, the deformation of Eudragit films without plasticizer on skin was greater than that on glass, suggesting a softer sample (Fig. 4.15a). The elastic modulus of this sample, using the Hertzian model, was therefore lower (0.40 ± 0.14

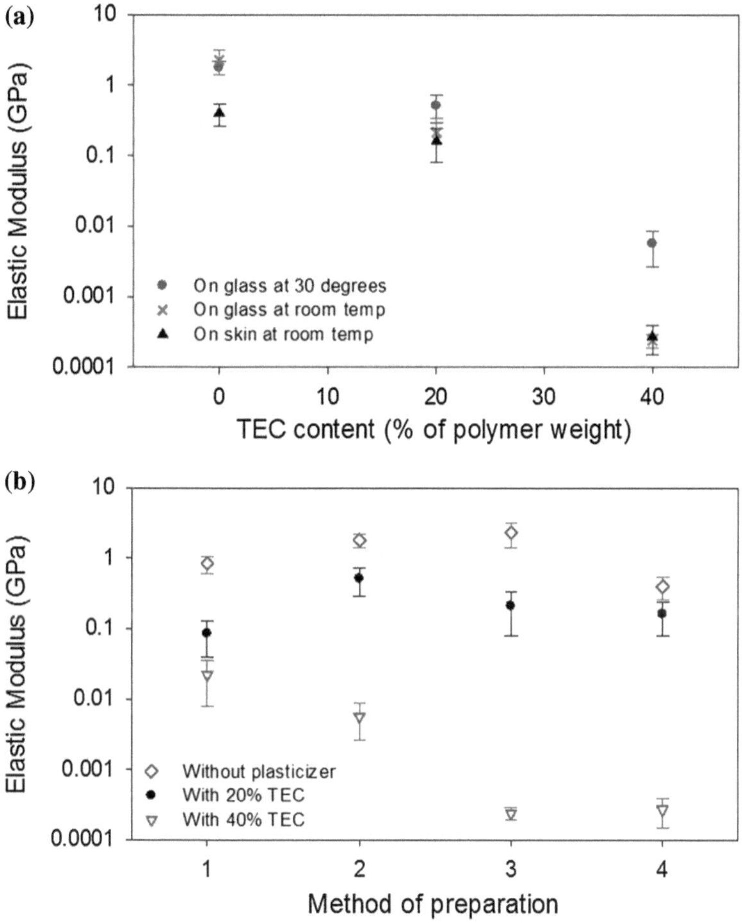

Fig. 4.16 Hertzian elastic moduli of Eudragit films (n = 8, ± standard deviation), without plasticizer and with 20 and 40 % TEC, prepared by different methods: **a** dried for 6 h on glass, at 30 °C and at room temperature, and on skin, at room temperature; **b** *method 1* dried for 24 h on glass at 30 °C, *method 2* dried for 6 h on glass at 30 °C, *method 3* dried for 6 h on glass at room temperature, and *method 4* dried for 6 h on porcine skin at room temperature

GPa) than seen before (Fig. 4.16b), implying that the film had been softened in some way by its interaction with the skin.

The elastic moduli of Eudragit films with either 20 or 40 % TEC, prepared on glass and on skin at room temperature, were not significantly different. It follows that the effect of TEC on the elastic modulus of the film is greater than the effect of skin on that which has no plasticizer incorporated.

The temperature of film formation affects the rate at which solvent evaporates from the FFS. Ethanol evaporation from these films occurs more quickly, of course, at higher temperatures. It has previously been observed that Klucel LF films require longer times to form from ethanolic solutions when formed at 25 °C than at 30 °C (1.5 times longer) [23]. For the same drying time, a greater amount of ethanol will therefore be left in a film when formed at room temperature than at 30 °C. Solvent, like plasticizer, can reduce polymer-polymer contacts as the film forms, and more solvent molecules will therefore have a greater plasticizing effect and lower the elastic modulus, exactly as is seen for films containing 40 % TEC prepared at room temperature (relative to those formed at 30 °C) [3].

All indentations were performed on the surface of the film exposed to air. The properties of the bulk film and those at its interface with the skin may be different. While the former may be deduced from examination of the fracture surface [1, 5], assessment of film properties at its interface with the skin offers practical challenges that are difficult to overcome with AFM because the skin surface topography (its furrows and wrinkles) has a dimension much greater than the height range ($\leq$5.5 μm) of the instrument.

4.5 Summary and Conclusions

AFM imaging has been used to determine the change in topography of polymeric films, cast on glass slides and dried for 24 h at 30 °C, with increasing plasticizer (TEC) content. RMS roughness decreased with increasing plasticizer content for Eudragit films. A similar change in roughness was not observed for Klucel films with the same plasticiser content. The difference may be explained by the higher molecular weight of Klucel relative to that of Eudragit. AFM images revealed homogeneous films with no obvious phase separation or crystallization.

AFM indentation experiments characterised the viscous, elastic, plastic behaviour of the polymeric films. Viscosity was evaluated by varying the rate at which the probe indented the films and by the use of a delay at maximum load.

Elastic moduli of polymeric films were extracted from indentation loops using the Hertzian model, at small deformations, and an adapted Oyen & Cook model. Hertzian model-calculated elastic moduli were consistently lower than those based on the Oyen & Cook model. The range of elastic moduli determined under different indentation conditions, using both models, was less than the effect of plasticizer incorporation. Elastic moduli of polymeric films decreased with increasing TEC content, with the incorporation of 20 % plasticizer reducing this parameter to that of skin (and allowing, thereby, the likely potential for such films to flex when in situ on the surface of the skin without breaking).

Chemical mapping by Raman micro-spectroscopy of films incorporating TEC and BMV showed relatively uniform distributions of plasticizer and drug over the examined area, implying that no phase separation or crystallization occurred. Mapping the peak frequency of the signal from BMV, revealed that the drug was dissolved in the polymeric films and that it was more soluble in Eudragit than in Klucel.

The impact of film preparation method (glass versus skin substrate, room temperature versus 30 °C, and 6 versus 24 h drying time) was explored using Eudragit films. In the absence of plasticizer, films formed on skin had lower elastic moduli than those prepared on glass, suggesting a possible plasticizing effect from moisture in the tissue. However, this was not observed when 20 or 40 % TEC was incorporated because the effect of the plasticizer overwhelmed that provoked by skin moisture.

References

1. M.E. Lauer, O. Grassmann, M. Siam, J. Tardio, L. Jacob, S. Page, J.H. Kindt, A. Engel, J. Alsenz, Atomic force microscopy-based screening of drug-excipient miscibility and stability of solid dispersions. Pharm. Res. **28**(3), 572–584 (2011)
2. Y.T.A. Tumer, C.J. Roberts, M.C. Davies, Scanning probe microscopy in the field of drug delivery. Adv. Drug Delivery Rev. **59**(14), 1453–1473 (2007)
3. M.E. Aulton, M.H. Abdulrazzak, J.E. Hogan, The mechanical-properties of hydroxypropyl-methylcellulose films derived from aqueous systems. 1. The influence of plasticizers. Drug Dev. Ind. Pharm. **7**(6), 649–668 (1981)
4. J.W. McGinity, L.A. Felton, *Aqueous polymeric coatings for pharmaceutical dosage forms*, 3rd edn., Drugs and the pharmaceutical sciences (Informa Healthcare, New York, 2008)
5. M.E. Lauer, M. Siam, J. Tardio, S. Page, J.H. Kindt, O. Grassmann, Rapid assessment of homogeneity and stability of amorphous solid dispersions by atomic force microscopy-from bench to batch. Pharm. Res. **30**(8), 2010–2022 (2013)
6. A. Danesh, X. Chen, M.C. Davies, C.J. Roberts, G.H.W. Sanders, S.J.B. Tendler, P.M. Williams, M.J. Wilkins, Polymorphic discrimination using atomic force microscopy: Distinguishing between two polymorphs of the drug cimetidine. Langmuir **16**(2), 866–870 (2000)
7. S. Ward, M. Perkins, J.X. Zhang, C.J. Roberts, C.E. Madden, S.Y. Luk, N. Patel, S.J. Ebbens, Identifying and mapping surface amorphous domains. Pharm. Res. **22**(7), 1195–1202 (2005)
8. R. Price, P.M. Young, Visualization of the crystallization of lactose from the amorphous state. J. Pharm. Sci. **93**(1), 155–164 (2004)
9. A.Q.J. Low, J. Parmentier, Y.M. Khong, C.C.E. Chai, T.Y. Tun, J.E. Berania, X.M. Liu, R. Gokhale, S.Y. Chan, Effect of type and ratio of solubilising polymer on characteristics of hot-melt extruded orodispersible films. Int. J. Pharm. **455**(1–2), 138–147 (2013)
10. F. Lecomte, J. Siepmann, M. Walther, R.J. MacRae, R. Bodmeier, Polymer blends used for the aqueous coating of solid dosage forms: importance of the type of plasticizer. J. Controlled Release **99**(1), 1–13 (2004)
11. S.Y. Lin, C.J. Lee, Y.Y. Lin, Drug-polymer interaction affecting the mechanical-properties, adhesion strength and release kinetics of piroxicam-loaded eudragit-e films plasticized with different plasticizers. J. Controlled Release **33**(3), 375–381 (1995)
12. C.B. Wu, J.W. McGinity, Non-traditional plasticization of polymeric films. Int. J. Pharm. **177**(1), 15–27 (1999)
13. J. Siepmann, R.A. Siegel, M.J. Rathbone, *Fundamentals and applications of controlled release drug delivery*, Advances in delivery science and technology (Controlled Release Society, Springer, New York, 2012)
14. M.L. Oyen, R.R. Cook, A practical guide for analysis of nanoindentation data. J. Mech. Behav. Biomed. Mater. **2**(4), 396–407 (2009)
15. M.L. Oyen, R.F. Cook, Load-displacement behavior during sharp indentation of viscous-elastic-plastic materials. J. Mater. Res. **18**(1), 139–150 (2003)
16. D.C. Lin, D.I. Shreiber, E.K. Dimitriadis, F. Horkay, Spherical indentation of soft matter beyond the hertzian regime: numerical and experimental validation of hyperelastic models. Biomech. Model. Mechanobiol. **8**(5), 345–358 (2009)
17. J.D. Beard, R.H. Guy, S.N. Gordeev, Mechanical tomography of human corneocytes with a nanoneedle. J. Invest. Dermatol. **133**, 1565–1571 (2013)

18. M.A. Repka, J.W. McGinity, Physical-mechanical, moisture absorption and bioadhesive properties of hydroxypropylcellulose hot-melt extruded films. Biomaterials **21**(14), 1509–1517 (2000)
19. C. Padula, G. Colombo, S. Nicoli, P.L. Catellani, G. Massimo, P. Santi, Bioadhesive film for the transdermal delivery of lidocaine: in vitro and in vivo behavior. J. Controlled Release **88**(2), 277–285 (2003)
20. J.C. Gutierrez-Rocca, J.W. Mcginity, Influence of water-soluble and insoluble plasticizers on the physical and mechanical-properties of acrylic resin copolymers. Int. J. Pharm. **103**(3), 293–301 (1994)
21. K. Frederiksen, R.H. Guy, K. Petersson, Formulation considerations in the design of topical, polymeric film-forming systems for sustained drug delivery to the skin. Eur. J. Pharm. Biopharm. **91**, 9–15 (2015)
22. L.A. Wegiel, L.J. Mauer, K.J. Edgar, L.S. Taylor, Crystallization of amorphous solid dispersions of resveratrol during preparation and storage-impact of different polymers. J. Pharm. Sci. **102**(1), 171–184 (2013)
23. J. Bajdik, G. Regdon, T. Marek, I. Eros, K. Suvegh, K. Pintye-Hodi, The effect of the solvent on the film-forming parameters of hydroxypropyl-cellulose. Int. J. Pharm. **301**(1–2), 192–198 (2005)

Chapter 5
Lipid Incorporation into Topical Polymeric Films

As discussed in the previous chapter, triethyl citrate (TEC) is one of several compounds, which are commonly used to plasticize polymeric films (others include dibutyl sebacate and tributyl citrate [1, 2]). The type of plasticizer incorporated into the films has been shown affect drug release [3–5]. Previous investigations of Eudragit and Klucel films reported increased release with increasing plasticizer lipophilicity, which prompted the assessment of medium-chain triglyceride (MCT) as a plasticizer [3]. MCT, containing fatty acids with 6–12 carbon atoms, are well-known excipients in dermatologic formulations [6]. Initially developed for the treatment of disorders of lipid absorption, MCT are now also used as emulsifiers and to improve the bioavailability of poorly water-soluble drugs [7–10].

This chapter describes the use of a similar approach to that used for films incorporating TEC as a plasticizer (Chap. 4). AFM imaging was used to determine the homogeneity of films formed when MCT was incorporated into Eudragit and Klucel FFS. Based on the observation of two-phase systems in both polymer films, tapping mode imaging parameters were varied to reveal the softer nature of the formed inclusions within a stiffer surrounding film. No drug crystallization was observed in either of the phases present in the films. The elastic moduli of these heterogeneous films, determined using AFM nanoindentation, varied from the moduli of polymeric films without MCT to factors of 3 and 4 times lower within the inclusions.

Raman micro-spectroscopy was carried out to determine the distribution of the polymers, MCT and the drug substance within the two phases observed. MCT was mostly confined within the inclusions (with a higher proportion being confined in Klucel films) and BMV was more evenly distributed throughout the films, but with a higher proportion found outside the inclusions. The physical state of BMV in the films was mapped using the shift of a characteristic spectral peak. The solubility

H. Garvie-Cook, *Novel (Trans)dermal Drug Delivery Strategies*,
Springer Theses, DOI 10.1007/978-3-319-28901-4_5

of the drug was approximately constant across the mapped area of Eudragit films incorporating MCT. In Klucel films, the inclusions provided an environment in which the drug was more soluble.

The observed phase separation implies that MCT does not act as a conventional plasticizer, as it was not distributed evenly throughout the polymeric film. Films incorporating MCT plasticizer did, however, show an enhancement in BMV release, compared to films without MCT. The mechanism behind this enhancement is discussed.

5.1 AFM Imaging

Images of the topography of films cast on cleaned glass slides at 30 °C were acquired using tapping mode AFM (Fig. 5.1). Incorporation of MCT significantly changed the topography of the polymeric films and revealed a two-phase structure (Fig. 5.1b, d). In Eudragit films, regions (from now on referred to as inclusions) were observed which appear to dip into the sample surface (Fig. 5.1c). The diameter of these inclusions, as calculated using images of 6 Eudragit films prepared in the same way, was 0.8 ± 0.2 μm. These inclusions were also observed in Klucel films incorporating MCT, but appear raised in Fig. 5.1d. The diameter of the inclusions formed in Klucel

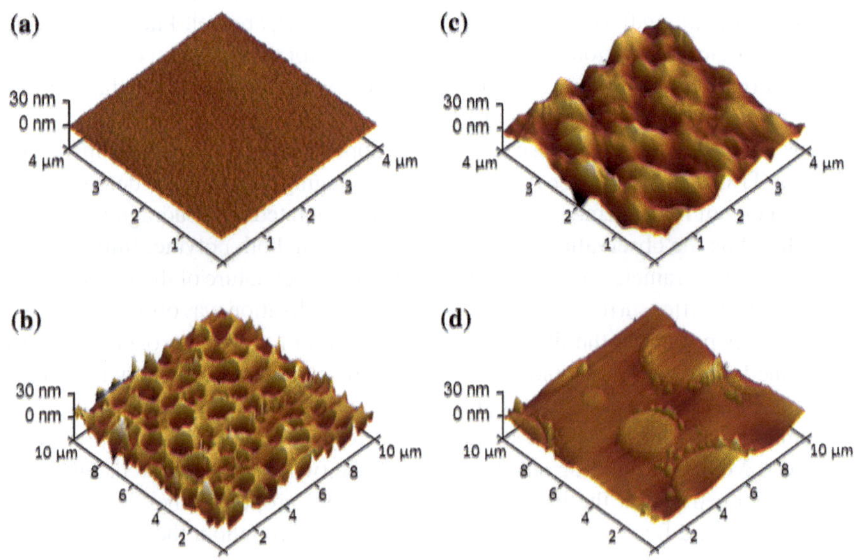

Fig. 5.1 3D AFM images of Eudragit (**a** and **b**) and Klucel (**c** and **d**) FFS, cast onto glass microscope slides, with (**b** and **d**) and without (**a** and **c**) 20 % w/w MCT incorporated. The scan size was $4 \times 4\ \mu m^2$ for the films without MCT and $10 \times 10\ \mu m^2$ for those with

films was larger, 1.9 ± 0.9 μm. The percentage areas that these inclusions cover were 22 ± 17 % and 33 ± 3 % for Eudragit and Klucel films, respectively. No drug crystals were observed in these images, suggesting that the drug is molecularly dispersed in both the inclusions and the surrounding areas.

The presence of these inclusions implies that MCT does not act as an effective plasticizer; it is not incorporated evenly and has not therefore had a homogeneous effect on the polymer [11]. This can be compared to the homogeneous appearance of Eudragit and Klucel films when a well-known plasticizer, TEC, was incorporated (Fig. 4.1). The MCT-containing films may now contain weak points, a result of these distinct regions, where fracture is more likely [10]. The drug delivery of the films may also be affected by the different affinities of the drug to the two phases and also by the presence of the inclusions, which may alter drug diffusivity [12].

Tapping mode AFM, as opposed to contact mode, is commonly used for soft samples, such as polymers and biological specimens, because the damage induced by the contact between the AFM tip and the sample is reduced [13]. The images acquired in tapping mode are not, however, purely topographical and the relation between the observed features and the actual surface topography must be carefully considered. To assess further the nature of the observed inclusions, tapping mode imaging parameters were varied.

5.1.1 Interpretation of Tapping Mode Height

In tapping mode, the cantilever is forced to oscillate vertically at a frequency close to its resonance frequency and the probe tip makes brief contact with the sample surface in each oscillation cycle [13]. The perturbation of the oscillation from this contact and from tip-sample interactions, which are the result of attractive Van der Waals and capillary forces, and the repulsive indentation force [13] (amongst others), cause the amplitude and resonance frequency to change. An increase in sample height due to a purely topographical feature will result in a decrease in amplitude of the cantilever oscillation. A feedback system then increases the height of the cantilever above the surface until the oscillation of the cantilever returns to its original amplitude, resulting in a change in height in the image.

A region composed of a material with a more viscous and/or elastic nature than its surroundings causes greater damping of the AFM oscillation. Over these regions, the height of the cantilever above the sample is raised to maintain the same amplitude of oscillation, resulting in an increase in height in the image even if the actual height of the sample has not changed. A change in sample hydrophobicity also has an effect on the tip-sample interaction. Under ambient conditions, a contamination layer, mainly composed of water, is present on all surfaces [14]. The capillary force from this layer attracts the AFM probe tip to the sample surface. Regions of greater hydrophilicity will have thicker contamination layers, resulting in greater oscillation damping and therefore higher apparent topography [13].

AFM images of Eudragit and Klucel films incorporating MCT showed depressed and raised inclusions, respectively (Fig. 5.1b, d). To understand the nature of these inclusions and determine whether the features seen were purely topographical or affected by mechanical heterogeneity, the set-point used for tapping mode imaging was varied. The amplitude set-point is the sustained amplitude of cantilever oscillation. A feedback system maintains this amplitude by varying the height of the AFM cantilever above the surface of the sample according to the change in the oscillation amplitude of the cantilever due to the above-mentioned factors.

AFM images of Eudragit and Klucel films with 20 % MCT were acquired at three set-points (Fig. 5.2). The height profiles along a line of the images, as indicated in Fig. 5.2g, revealed differences in the topography of the images depending on the set-point used. The structures observed were therefore not purely topographical and were affected by the imaging process. When the highest set-point (1.338 V) was used to acquire an image of Eudragit with 20 % MCT (Fig. 5.2a), topographical features with the lowest height differences were observed. The inclusions appeared less than ~10 nm deep and the AFM image showed the least contrast between the inclusions and the surrounding areas of the film. In the equivalent image of Klucel with 20 % MCT (Fig. 5.1d), the inclusions appeared raised from the surrounding areas and have heights of approximately 6 nm. As the set-point decreased, the inclusions appeared deeper in Eudragit with 20 % MCT, reaching depths of 80 nm (Fig. 5.2g). Inclusions appeared depressed into the sample surface in Klucel with 20 % MCT, with the lowest set-point resulting in inclusions with apparent depths of approximately 10 nm.

The apparent topography observed is likely to have arisen from regional variations in mechanical properties. The cantilever is driven to oscillate freely at a given amplitude and frequency at the start of the imaging process. The set-point amplitude is less than this free amplitude as the oscillation of the cantilever is damped as it approaches the surface and the tip-sample interaction increases. A lower set-point (greater difference to the free amplitude than a higher set-point) is maintained by greater damping of the oscillation via the tip-sample interaction. The force between the tip and the sample is therefore greater at lower set-points. At a given set-point, areas of the sample which have lower elastic moduli and a more viscous nature will deform more under the force of the tip-sample contact. When the set-point was decreased, the tip-sample force increased and deformation of the (softer) inclusions increased, resulting in the appearance of deeper features in AFM images (Fig. 5.2).

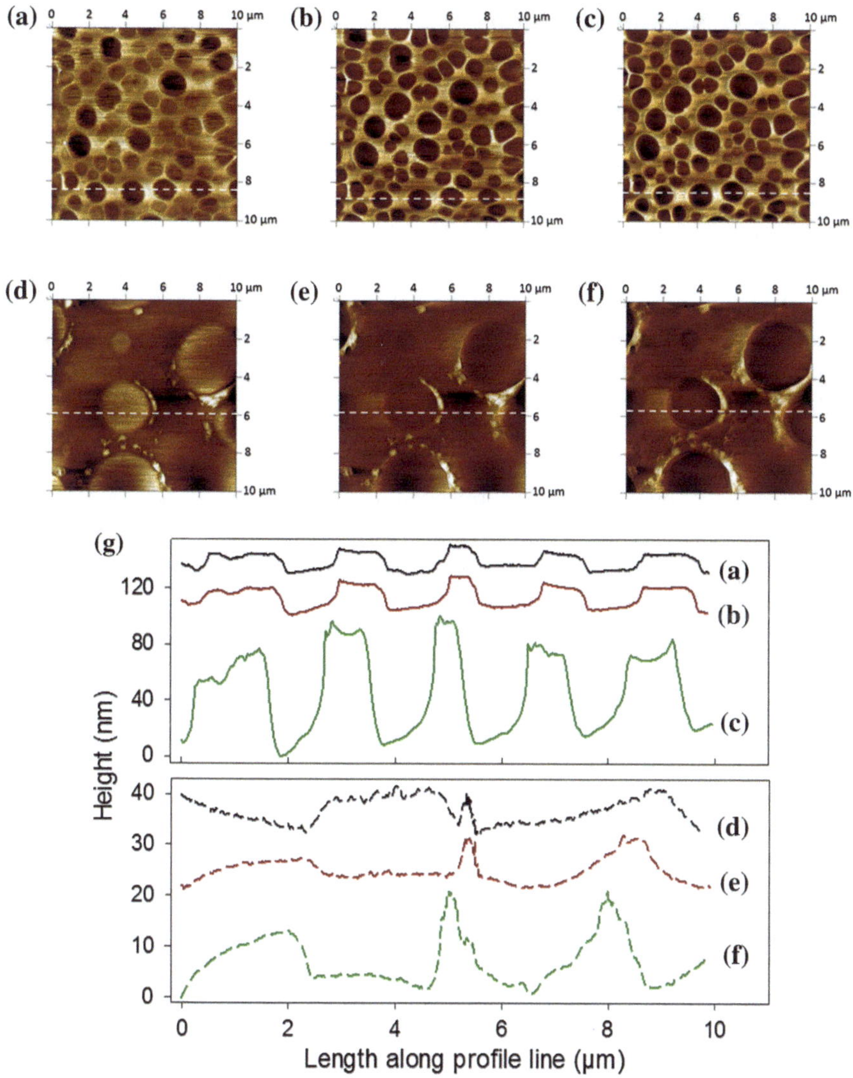

Fig. 5.2 2D AFM tapping mode images of (**a**–**c**) Eudragit and (**d**–**f**) Klucel films incorporating 20 % MCT. From *left* to *right*, the set-point for imaging decreased from (**a**) 1.338 to (**b**) 1.288 to (**c**) 1.138 V for Eudragit films, and (**d**) 2.281 to (**e**) 2.231 to (**f**) 2.136 V for Klucel films. *Dashed white lines* indicate the line of the image used to obtain height profiles (**g**) as a function of distance along this line for the three decreasing set-points used to acquire images of each film

5.2 AFM Nanoindentation

5.2.1 *Deformation Behaviour*

AFM nanoindentation of films incorporating MCT was carried out to compare their mechanical properties to those of films without MCT. Examples of the indentation loops of Eudragit and Klucel films incorporating MCT, as load is applied to the surface using an AFM probe tip, are shown in Fig. 5.3. Due to the softer nature of the inclusions, as predicted from Sect. 5.1.1, indents that require the lowest loads for the most deformation are considered to be from inclusions. The areas surrounding the inclusions were thought to be stiffer. Areas where the greatest loads were required to deform the film the least were therefore considered to be regions of the films that surrounded the inclusions. Nanoindentation of the areas surrounding the inclusions in both Eudragit and Klucel films incorporating MCT revealed similar indentation behaviour to that observed in the corresponding films without MCT (Fig. 5.3).

5.2.2 *Elastic Moduli*

Elastic moduli of the samples were determined using the Oyen & Cook model, as described in Sect. 3.1.3.4. An example of the fit of this model to the indentation of Eudragit with 20 % MCT is in Fig. 5.4, showing good agreement between the model and the experimental data.

Indentation was performed at intervals of 250 or 500 nm along a line across the surface of the film. Examples of elastic modulus as a function of distance along this line, for both Eudragit and Klucel films incorporating MCT, are shown in Fig. 5.5. The elastic modulus values, determined for each indent performed, varied from the value of films without MCT to minimum values of 0.29 and 0.31 GPa for Eudragit and Klucel films, respectively. The values of elastic modulus across the lines of indents showed continuous variation between the maximum and minimum values. Regions along the lines of indents with consecutively low elastic moduli were thought to be inclusions. The length scale of these regions corresponded to the dimensions of the inclusions. The average values of elastic moduli of the inclusions were 0.44 ± 0.12 GPa (n = 6, $\pm$SD) and 0.39 ± 0.09 GPa (n = 5, $\pm$SD) for Eudragit and Klucel films, respectively. These averages are not significantly different ($p>0.05$, unpaired t-test), suggesting that the compositions of the inclusions in Eudragit and Klucel films are similar. The variation in elastic modulus is thought to be continuous (as opposed to 'binary') due to confinement effects, where the softer and stiffer segments influence each other at and near their interfaces [15].

Although the elastic moduli of the inclusions in Eudragit and Klucel films were numerically similar, their deformation behaviours differed. The indentation of inclusions, specifically the hysteresis between the approach and retract data, and the non-linearity during loading and unloading, revealed their softer and more viscous nature than the surrounding areas of the films. Inclusions in Eudragit films deformed

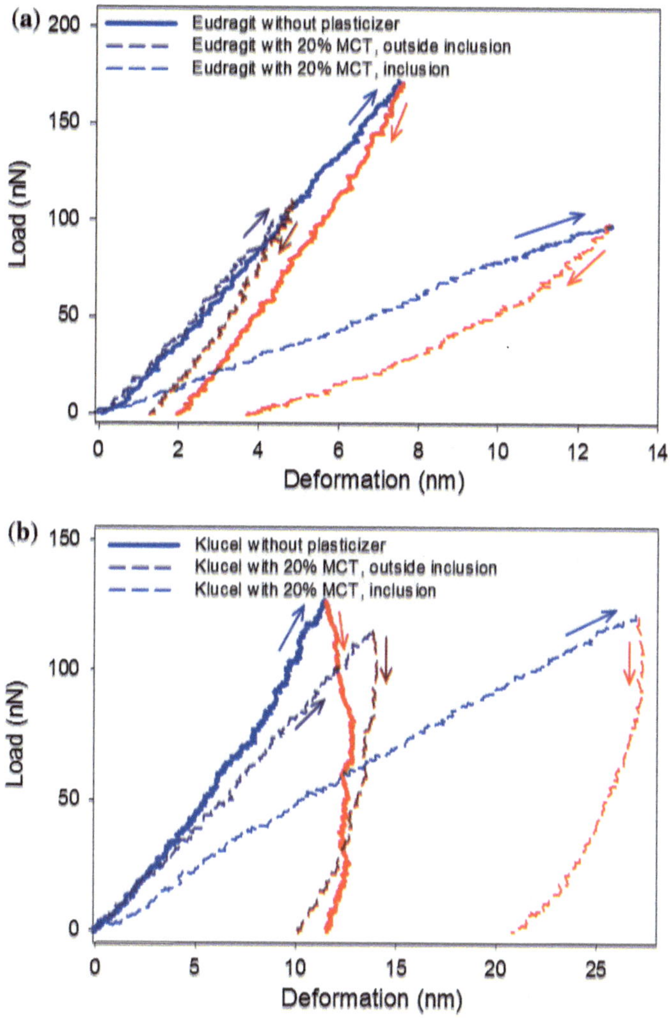

Fig. 5.3 Load as a function of deformation during indentation of (**a**) Eudragit ± MCT, and (**b**) Klucel ± MCT. Examples of indents from areas from within and from outside the observed inclusions are shown

more under a given load than films without MCT, but the nature of the deformation of the two was similar. Indentation of inclusions in Klucel films revealed their similar viscous-elastic-plastic nature to the surrounding areas and to films with no MCT. The inclusions, however, deformed more under a given load due to their softer nature. The inclusions were therefore influenced by the polymer in which they were incorporated and they were likely to contain polymer as well as MCT. To confirm the distribution of the constituents of polymeric films incorporating MCT relative to the topographical features observed, Raman chemical mapping was performed (Sect. 5.3).

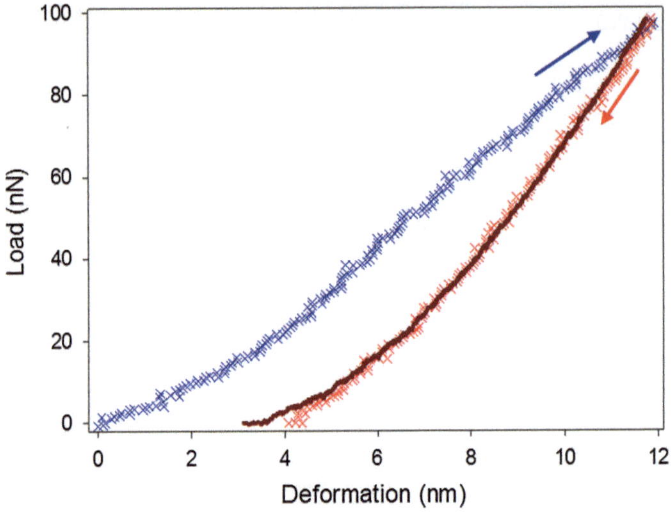

Fig. 5.4 The Oyen & Cook model fit (*solid dark red line*) to the experimentally measured load as a function of deformation (*crosses*) during indentation of Eudragit with 20 % MCT

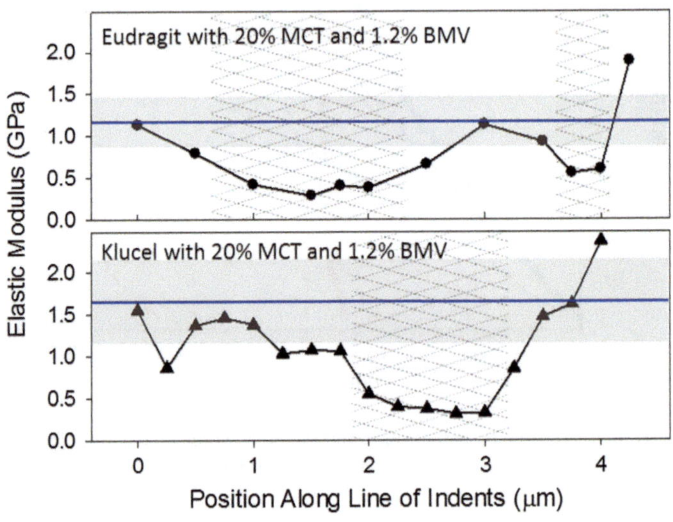

Fig. 5.5 Elastic modulus as a function of position along a line of indents in Eudragit and Klucel films containing 20 % MCT. Areas shaded with crosshatching are considered to be inclusions. The elastic moduli of polymeric films without MCT are shown as *horizontal blue lines* with *grey shading* indicating the corresponding standard deviations

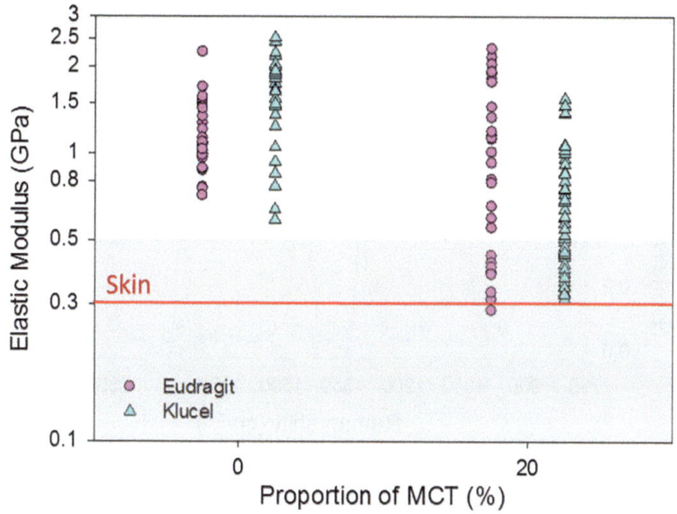

Fig. 5.6 Elastic moduli of Eudragit and Klucel films incorporating 0 and 20 % MCT. The elastic modulus of skin [16] is shown as a *horizontal red line* for comparison, with the standard deviation shown in *grey*. The data were acquired using a minimum of 8 indents on 3 films of each composition

All values of elastic modulus determined for Eudragit and Klucel films with and without MCT are shown in Fig. 5.6 (a minimum of 8 indents on 3 samples of each film were performed). Due to the different compositions of the inclusions and their surroundings, a large spread of results was observed for films incorporating MCT. Calculating an average elastic modulus is not representative of films incorporating MCT due to their heterogeneous structure. Elastic moduli of inclusions approached the elastic modulus of skin [16]. However, the comparison of the mechanical properties of these films with skin is more complex in this case, compared to films plasticized with TEC (Sect. 4.2.2.2). The stiffer regions of the films are more likely to break contact than the inclusions when the film is on the skin, which constantly stretches and flexes [17]. These films, however, show more rapid release compared to films plasticized with TEC, so are advantageous in this sense. This is further discussed in Sect. 5.4.

5.3 Raman Micro-Spectroscopy

5.3.1 MCT Characteristic Spectral Features

A Raman spectrum of MCT was acquired to determine characteristic peaks by which it could be identified in spectroscopic maps of Eudragit and Klucel films into which it was incorporated (Fig. 5.7). Characteristic spectral peaks were identified by

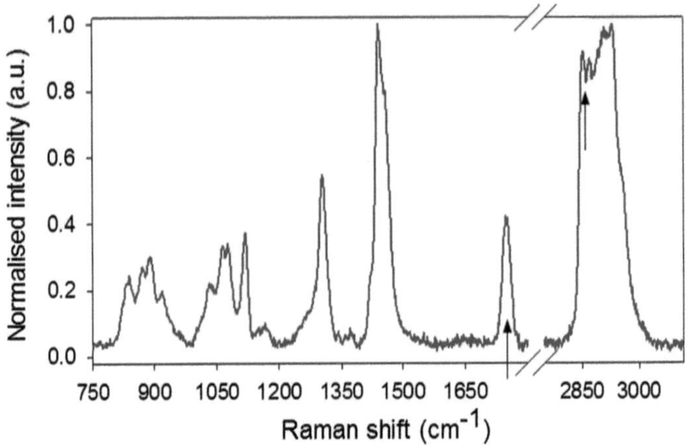

Fig. 5.7 Raman spectrum of MCT, with *arrows* indicating the positions of spectral features used for identifying the compound in Raman chemical maps. This spectrum has been normalised within the ranges shown (750–1800 cm^{-1} and 2750–3120 cm^{-1})

comparing this spectrum with the spectra of Eudragit, Klucel and BMV (Fig. 4.11). The spectrum of MCT showed a spectral peak at 2880 cm^{-1}, a frequency at which Eudragit showed only low intensity features. The intensity of this peak in Raman maps of Eudragit films incorporating MCT was therefore used to determine the distribution of the former within these films. In Klucel, another characteristic peak at 1745 cm^{-1} was used to map MCT (this peak was not used to map MCT in Eudragit films as it overlapped with features specific to Eudragit). The BMV-specific peak around 1666 cm^{-1}, as identified in Sect. 4.3.1, was used to map the distribution of the drug across the films. The signal from the drug at higher wavenumbers (2800–3100 cm^{-1}) was of insufficient intensity to be useful.

For Raman chemical mapping, films incorporating MCT were formed from solution at room temperature, as opposed to the 30 °C used for AFM samples. The inclusions formed at this lower temperature were larger (3.9 ± 1.0 μm and 3.5 ± 1.0 μm for Eudragit and Klucel films, respectively), facilitating the use of (lower resolution) Raman micro-spectroscopy to map the constituents of the films within the inclusions formed.

5.3.2 Mapping Concentration

Maps of the distributions of MCT and BMV in Eudragit and Klucel films are in Fig. 5.8 and spectra from bright and dark pixels are shown in Fig. 5.8f. Brighter pixels occur when the intensity of the Raman signal is higher, and hence there is a greater concentration of the corresponding molecules. MCT maps of the two polymer films (Fig. 5.8a, c) show that there was a greater concentration of MCT within the

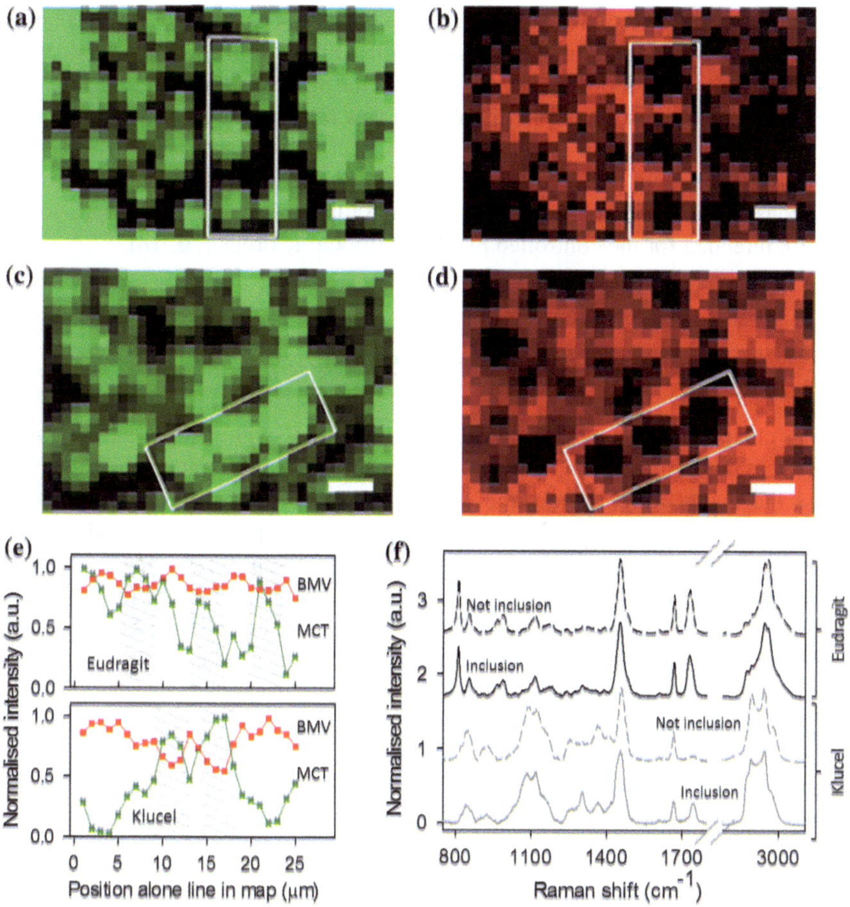

Fig. 5.8 Raman chemical maps of the distributions of MCT and BMV in polymer films. The distributions of MCT in Eudragit and Klucel are shown in (**a**) and (**b**), respectively; the corresponding distributions of BMV are in (**c**) and (**d**). *Brightly coloured green* (MCT) and *red* (BMV) pixels indicate higher concentrations; *black pixels* reflect areas of lower concentrations. *White boxes* in the maps highlight areas with clear differentiation in chemical levels. *Scale bars* represent 5 μm. The variations of Raman signals from MCT and BMV along the lines drawn across the maps are shown in (**e**), with the positions of inclusions indicated by *diagonal shading*. Example Raman spectra from within and outside inclusions are shown in (**f**). Spectra have been normalised within the ranges shown (750–1800 cm^{-1} and 2750–3120 cm^{-1}) and offset to facilitate visualization

inclusions than outside. The intensity difference (and therefore concentration difference) across a representative line of the Raman map is shown in Fig. 5.8e. Percentages were calculated by taking into account the underlying intensity of the polymer signal at the positions of MCT characteristic peaks. MCT is found throughout the film but the differences in concentration between a bright pixel and a dark pixel (corresponding to the highest and lowest concentrations within the mapped area) were 87 % in

Eudragit films and 96 % in Klucel films. A greater amount of the MCT in Klucel films was therefore found in the inclusions than in the Eudragit films. BMV showed the opposite distribution: more drug was located in the areas surrounding the inclusions (Fig. 5.8b, d) but the variation in distribution was less pronounced, with maximum differences of 24 and 44 % in Eudragit and Klucel films, respectively. The distribution of the polymers reflected that of the drug; there was less polymer within the inclusions. The segregation of MCT in Eudragit films has been observed previously in those intended for the controlled release of lipophilic drugs [10, 18].

5.3.3 *Mapping the Physical State of the Drug*

Raman chemical maps of the physical state of BMV within films incorporating MCT were determined from the shift in the position of the characteristic drug peak across the mapped area. The same method was used for the films incorporating TEC, as described in Sect. 4.3.3. In Eudragit films, the position of the drug peak was essentially constant at $1668.3 \pm 0.4\,\text{cm}^{-1}$ across the film. The BMV drug peak in Klucel films varied from $1667.43 \pm 0.15\,\text{cm}^{-1}$ in the inclusions to $1666.06 \pm 0.03\,\text{cm}^{-1}$ in the surrounding areas (Fig. 5.9). Bright pixels in Fig. 5.9a show areas where the BMV peak has a higher Raman shift; dark pixels indicate where the BMV peak occurs at lower frequency. The colour scale in this case has no relation to the concentration of the compound.

The positions of the BMV peaks in Eudragit and Klucel films (in the latter case, both within and outside of the inclusions) differ from one other. This agrees with the AFM indentation behaviour (Fig. 5.3), which suggested that the properties of the

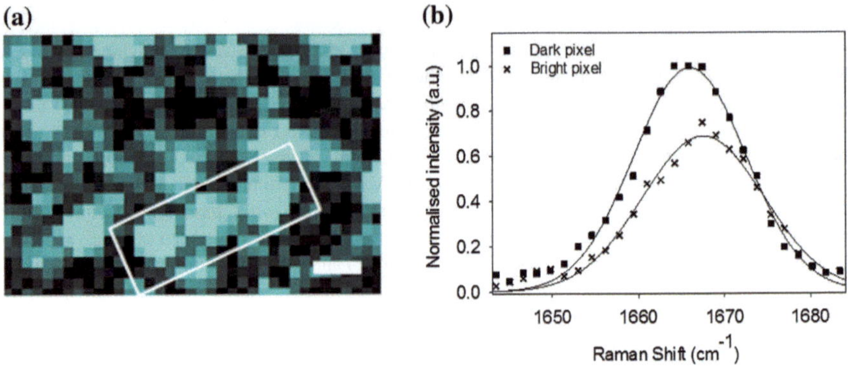

Fig. 5.9 **a** Raman map of the peak position of the BMV signal from a Klucel film: the brightest pixels originate from an average peak position of $1667.6\,\text{cm}^{-1}$, while the *dark pixels* correspond to $1665.8\,\text{cm}^{-1}$. *Scale bar* represents 5 μm. (**b**) Representative BMV spectra from *bright* and *dark pixels* in the Raman map of a Klucel film. Spectra have been normalised such that the maximum intensity of the peak shown across the mapped area was 1

inclusions were influenced by the polymer in which they were found. The polymer is therefore having an effect on the solubility of the drug within the inclusions.

The areas of bright intensity in the maps of BMV peak position in Klucel films incorporating MCT (the areas of higher peak position) corresponded to areas of lower intensity in the drug concentration map (Fig. 5.8d). This implies that there is a lower concentration of the drug substance (intensity of the Raman peak is lower) in the inclusions and that the drug is solubilised in these areas (position of the Raman peak is higher) (Fig. 5.9b). In Eudragit films, the drug appears to be equally soluble in all areas of the film.

5.4 Drug Release Mechanism

In vitro BMV release studies were conducted, using the same method as that described in [3], to determine the effect of MCT incorporation (at 20 %) into Eudragit and Klucel films [19]. For both polymers, sustained BMV release over 72 h was observed and a significant enhancement in release was apparent from about 6 h into the experiment, in the presence of MCT (Fig. 5.10). The rate of release was observed to increase from 0.45 to 2.19 $\mu g\ cm^{-2}\ h^{-\frac{1}{2}}$ in Eudragit films, and from 0.58 to 1.56 $\mu g\ cm^{-2}\ h^{-1}$ in Klucel films, when MCT was incorporated. After 72 h, drug release from Eudragit and Klucel was enhanced by factors of 4.5 and 2.5, respectively.

These results suggest that the observed enhancement in release with the incorporation of MCT can be attributed to the presence of inclusions. These inclusions were observed in AFM images of the films (Fig. 5.1). By varying the imaging conditions, namely the tapping mode set-point amplitude, it was observed that the inclusions were composed of a material softer than their surroundings (Fig. 5.2). This was confirmed with nanoindentation measurements (Figs. 5.5 and 5.6); the elastic moduli of the inclusions were approximately 3 and 4 times lower than the surrounding

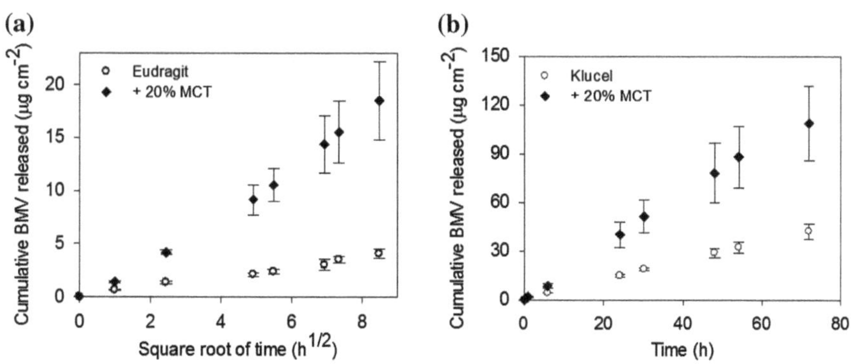

Fig. 5.10 In vitro release of BMV from (**a**) Eudragit, and (**b**) Klucel films with and without MCT (mean ± standard deviation, n = 3). Modified from [19]

material in Eudragit and Klucel films, respectively. The indentation behaviours of the inclusions were dependent on the polymer that contained them (Fig. 5.3).

Raman chemical mapping showed that a greater proportion of the MCT within the films was contained within the inclusions (Fig. 5.8). Drug was distributed more evenly than MCT but showed higher concentrations in the areas surrounding the inclusions than within them (Fig. 5.8). In Klucel films, the drug was in a solubilised form within the inclusions (Fig. 5.9a).

In Eudragit films with and without MCT, the peak of the Raman signal originating from BMV was approximately the same across the surface of the films (1668.3 $\pm$ 0.4 cm^{-1}). The incorporation of MCT did not, therefore, change the solubility of the drug in this polymer film significantly. The enhancement in drug release observed may originate from the lower elastic modulus and more viscous nature of the inclusions. By incorporating MCT, a larger area of the film (which would be in contact with the skin or the silicone membrane as used in the release studies [3]) has a lower elastic modulus and a lower viscosity. Regions of lower elastic moduli and lower viscosity provide an environment in which the drug is more highly mobile. More BMV therefore diffuses out of the film in a given time than out of films without this phase separation (i.e., without MCT). The fractional area of the inclusions in Eudragit was not significantly different from that in Klucel but the enhancement in release after 72 h was greater. The enhancement in release cannot, therefore, solely originate from, for example, a larger total area in which the drug is more highly mobile.

The Raman peak position of BMV in Klucel films incorporating MCT shifted according to its environment, whether in the MCT-rich inclusions or the surrounding areas. The drug was more soluble within the inclusions, where a higher concentration MCT, and lower concentrations of polymer and the drug, were present. The solubility of BMV outside the inclusions was indistinguishable from that in Klucel films without MCT, as determined from Raman peak position measurements. Drug release has been enhanced in this case, therefore, due to the softer nature of the inclusions and the role of the inclusions in providing an environment in which the drug is more soluble. The lower enhancement in drug release, compared to the enhancement when MCT is incorporated into Eudragit films, may arise from the lower solubility of BMV in Klucel.

5.5 Summary and Conclusions

A two-phase film, observed using AFM imaging, was formed upon solvent evaporation from FFS incorporating MCT. Tapping mode imaging and AFM nanoindentation revealed the softer and more viscous nature of the inclusions observed. The distributions of MCT and BMV relative to the observed inclusions were mapped using Raman micro-spectroscopy and more MCT was found within the inclusions. The drug was distributed more evenly than MCT but a higher concentration of BMV was observed in the areas of the films surrounding the inclusions. The physical state of the drug was mapped using the position of the Raman spectral peak specific to

BMV. The drug was equally soluble in both phases in Eudragit films. BMV showed a greater solubility in the inclusions in Klucel films with MCT, but remained dissolved throughout the film. The enhanced rate of release observed for films incorporating MCT was attributed to the presence of the inclusions.

The incorporation of MCT into topical polymeric films for drug delivery therefore has advantages and disadvantages. The heterogeneity, resulting from the formation of MCT-rich inclusions, may result in fracture of the films when the areas surrounding the inclusions are stretched [10]. The presence of these inclusions does, however, enhance the rate of release from these films. These films would therefore be more useful when a high rate of release and low substantivity of the film on the skin is required. The substantivity of the films could be increased by incorporating both TEC and MCT, one as a conventional plasticizer, the other as a release-enhancer.

References

1. F. Lecomte, J. Siepmann, M. Walther, R.J. MacRae, R. Bodmeier, Polymer blends used for the aqueous coating of solid dosage forms: importance of the type of plasticizer. J. Control. Release **99**(1), 1–13 (2004)
2. R. Bodmeier, O. Paeratakul, The distribution of plasticizers between aqueous and polymer phases in aqueous colloidal polymer dispersions. Int. J. Pharm. **103**(1), 47–54 (1994)
3. K. Frederiksen, R.H. Guy, K. Petersson, Formulation considerations in the design of topical, polymeric film-forming systems for sustained drug delivery to the skin. Eur. J. Pharm. Biopharm. **91**, 9–15 (2015)
4. C.B. Wu, J.W. McGinity, Non-traditional plasticization of polymeric films. Int. J. Pharm. **177**(1), 15–27 (1999)
5. J.C. Gutierrez-Rocca, J.W. Mcginity, Influence of water-soluble and insoluble plasticizers on the physical and mechanical-properties of acrylic resin copolymers. Int. J. Pharm. **103**(3), 293–301 (1994)
6. K.A. Traul, A. Driedger, D.L. Ingle, D. Nakhasi, Review of the toxicologic properties of medium-chain triglycerides. Food Chem. Toxicol. **38**(1), 79–98 (2000)
7. A.J. Humberstone, W.N. Charman, Lipid-based vehicles for the oral delivery of poorly water soluble drugs. Adv. Drug Deliv. Rev. **25**(1), 103–128 (1997)
8. C.J.H. Porter, A.M. Kaukonen, B.J. Boyd, G.A. Edwards, W.N. Charman, Susceptibility to lipase-mediated digestion reduces the oral bioavailability of danazol after administration as a medium-chain lipid-based microemulsion formulation. Pharm. Res. **21**(8), 1405–1412 (2004)
9. S. Laube, M.G. Davies, L. Prais, I.S. Foulds, Allergic contact dermatitis from medium-chain triglycerides in a moisturizing lotion. Contact Dermat. **47**(3), 171–171 (2002)
10. D.J. Lunter, R. Daniels, New film forming emulsions containing eudragit ne and/or rs 30d for sustained dermal delivery of nonivamide. Eur. J. Pharm. Biopharm. **82**(2), 291–298 (2012)
11. J.W. McGinity, L.A. Felton, *Aqueous Polymeric Coatings for Pharmaceutical Dosage Forms*, 3rd edn., Drugs and the Pharmaceutical Sciences (Informa Healthcare, New York, 2008)
12. J. Siepmann, R.A. Siegel, M.J. Rathbone, *Fundamentals and Applications of Controlled Release Drug Delivery*, Advances in Delivery Science and Technology (Springer, Controlled Release Society, New York, 2012)
13. G. Bar, Y. Thomann, R. Brandsch, H.J. Cantow, M.H. Whangbo, Factors affecting the height and phase images in tapping mode atomic force microscopy. study of phase-separated polymer blends of poly(ethene-co-styrene) and poly(2,6-dimethyl-1,4-phenylene oxide). Langmuir **13**(14), 3807–3812 (1997)

14. B. Cappella, G. Dietler, Force-distance curves by atomic force microscopy. Surf. Sci. Rep. **34**(1–3), 1–104 (1999)
15. P. Schon, K. Bagdi, K. Molnar, P. Markus, B. Pukanszky, G.J. Vancso, Quantitative mapping of elastic moduli at the nanoscale in phase separated polyurethanes by afm. Eur. Polym. J. **47**(4), 692–698 (2011)
16. J.D. Beard, R.H. Guy, S.N. Gordeev, Mechanical tomography of human corneocytes with a nanoneedle. J. Investig. Dermatol. **133**, 1565–1571 (2013)
17. I.Z. Schroeder, P. Franke, U.F. Schaefer, C.M. Lehr, Development and characterization of film forming polymeric solutions for skin drug delivery. Eur. J. Pharm. Biopharm. **65**(1), 111–121 (2007)
18. D. Lunter, R. Daniels, In vitro skin permeation and penetration of nonivamide from novel film-forming emulsions. Skin Pharm. Physiol. **26**(3), 139–146 (2013)
19. H. Garvie-Cook, K. Frederiksen, K. Petersson, R.H. Guy, S.N. Gordeev, Biophysical elucidation of the mechanism of drug release enhancement from topical, polymeric film-forming systems. J. Control. Release **212**, 103–112 (2015)

Chapter 6
Laser Microporation of Skin

The percutaneous permeation of drugs across the skin is greatly enhanced when the stratum corneum (SC) is compromised by physical poration. The presence of pores allows even the transport of larger molecules, such as proteins, peptides and vaccines, across the skin [1, 2]. Exposure to laser radiation can be used to ablate the skin and form pores. The mechanism of ablation depends on the wavelength and pulse duration of the laser used.

Currently, devices available for laser microporation of the skin are used both for drug delivery and cosmetic applications [1, 3]. The pulse durations involved are in the microsecond range and ablate the skin photothermally; the water within the skin absorbs the laser light and vaporizes, causing ablation. However, the lasers required are expensive and their use is associated with side effects, such as skin irritation and a prolonged healing period [4, 5].

Ablation of skin using visible laser light has been less studied and requires a higher fluence for ablation and permeation enhancement compared to infrared (IR) and ultraviolet (UV) lasers [6, 7]. The use of visible light, however, is safer as unintentional exposure is easily observed and can therefore be controlled. Demonstrations of skin ablation using a femtosecond pulsed laser indicated that the mechanism involved plasma formation [8–10]. Histological examination of the skin after exposure to the laser beam revealed some tissue damage around the ablated pore but, due to its shorter pulse duration, femtosecond ablation has the potential to cause much less thermal perturbation than microsecond pulsed devices. Femtosecond ablation of porcine mammalian has therefore been investigated here, with a view to decreasing the fluence required for ablation and therefore limiting the extent of thermal damage.

Porcine skin has been porated using a fibre laser with a wavelength of 532 nm and pulse durations of approximately 300 fs. A dye was applied to the surface of the skin, enabling the use of lower laser power. The ability of this ablation technique to enhance drug permeation across the SC was investigated with caffeine as a model drug. At lower powers, when poration was facilitated by the ink on the surface of the skin, less thermal damage, as determined using Raman spectroscopy, was observed in the tissue surrounding the pores. The diameters of the pores could be varied by

H. Garvie-Cook, *Novel (Trans)dermal Drug Delivery Strategies*,
Springer Theses, DOI 10.1007/978-3-319-28901-4_6

changing the laser power and the distance between the skin and the fibre used to direct the light towards the skin. Greater enhancement in the permeation of caffeine was observed with increasing laser power. The use of both a fibre laser, and a fibre to direct the laser light towards the skin, allows for flexibility in the design of a portable device for laser microporation.

6.1 Pore Dimensions

Porcine skin was porated either without the application of an ink (uninked) or shortly after ink was applied to its surface (inked). The laser light was focussed on the skin using a lens (lens set-up) or directed to the skin using a fibre (fibre set-up).

6.1.1 Optical Microscopy of Laser Micropores

Examples of pores formed by focussing laser light onto inked and uninked skin using a lens are shown in Fig. 6.1. The appearance of burnt tissue in uninked skin was clear, in Fig. 6.1a, b, around the sides and at the bottom of the pore. The occurrence of tissue damage in inked skin was less obvious (Fig. 6.1e, f) and the bottom of the pores was similar in appearance to the undamaged skin surrounding the pore. The lack of obvious thermal damage may reflect the fact that the black ink obscures visualisation of black, burnt tissue. Alternatively, it may be the case that there really is less damage to the tissue when the skin is inked. The pore in inked skin shown in Fig. 6.1e was formed using a lower power than that in uninked skin (Fig. 6.1a) (0.7 W as opposed to 1.3 W) which should result in less thermal damage. The ink influences the mechanism of poration, causing in less damage to the surrounding tissue and is further discussed in Sect. 6.4.

Cross-sectional images (Fig. 6.1c–d and g–h) were obtained by slicing through the middle of pores using a surgical scalpel. The skin was then supported on its side (with the cut surface facing upwards) and examined with an optical microscope. The images of pores in uninked skin show burnt tissue at the top and sides of the pores. In contrast, tissue burn in inked skin was again less obvious.

Laser light was also directed to the skin using a hollow core photonic crystal fibre (PCF). This fibre was specifically designed to guide femtosecond pulsed light with a wavelength of 532 nm, as used in these investigations. Fibres with air cores allow light with such short pulses to be guided without distortion. In more standard, silica fibres, nonlinear effects would alter the pulse and the high peak powers may cause damage [11].

An example of a pore produced by directing laser light to the skin using a fibre is shown in Fig. 6.2. Pores in inked skin were similar in appearance to those produced using the lens set-up but lower powers were used, resulting in smaller pores. The

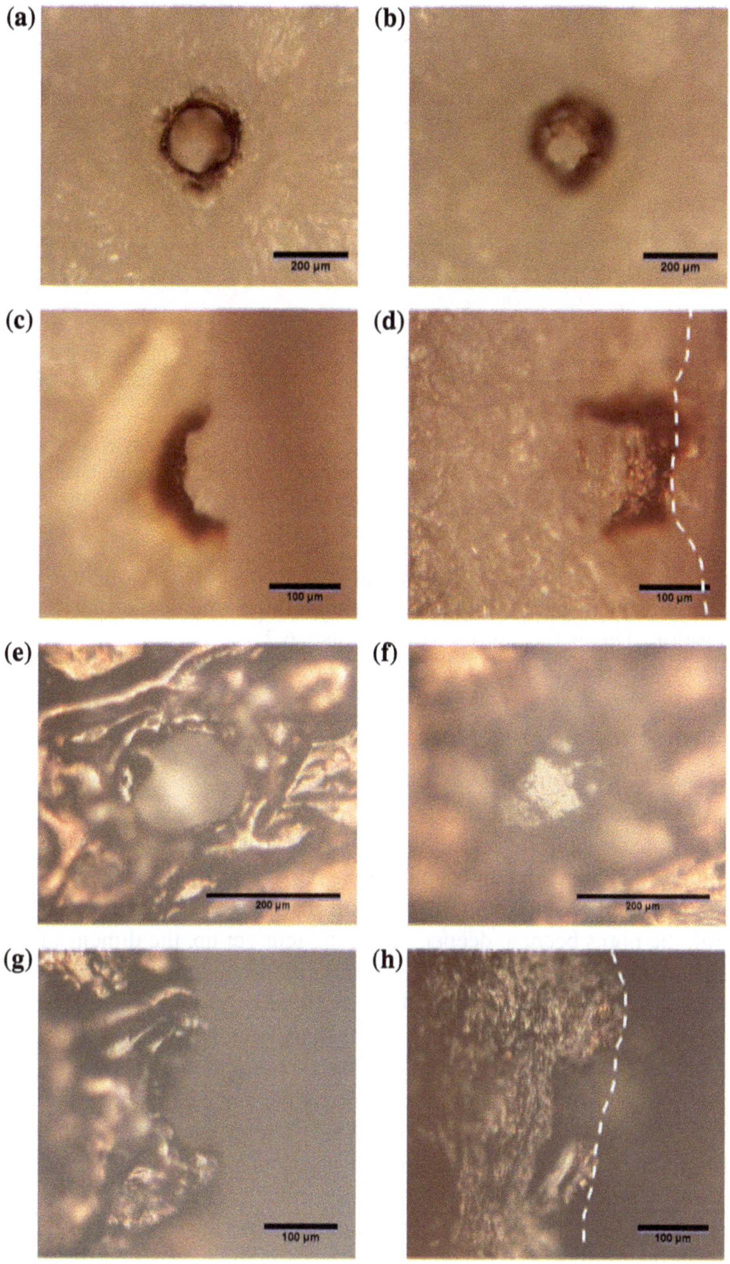

Fig. 6.1 Optical microscopy images of laser pores in uninked **a–d** and inked skin **e–h**, produced by focussing laser light onto the skin using a lens. The microscope was focussed on the *top* (**a** and **e**) and bottom (**b** and **d**) of the pores. Cross section images were taken from the top down (**c** and **g**) and from the side (**d** and **h**). *White dashed lines* mark the uppermost surface of the skin. Laser powers of 1.0 W (**a** and **b**), 1.3 W (**c** and **d**) and 0.7 W (**e–h**) were used. Each row of images is obtained from a different pore. The exposure time to the laser was 1 s

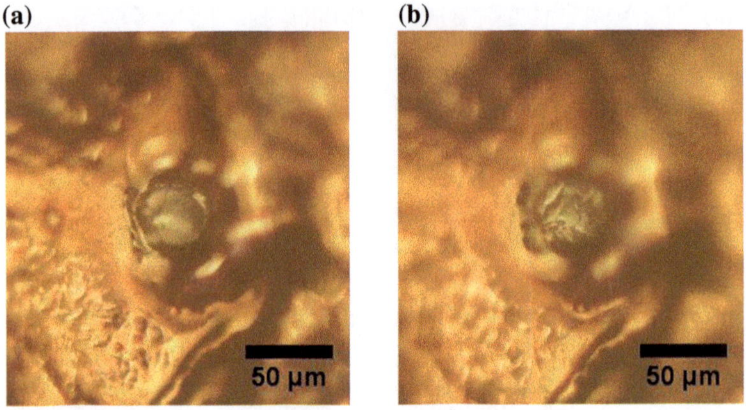

Fig. 6.2 Optical microscopy images of the *top* (**a**) and *bottom* (**b**) of a pore in inked skin, formed by directing laser light onto the skin using a photonic crystal fibre. Images are taken from the top down. A laser power of 0.1 W was used to porate the skin. The exposure time to the laser was 1 s

bottom of the pore (Fig. 6.2a) does not appear to be burnt. A method was established, using Raman spectroscopy, to determine the extent of thermal damage after microporation. This is further described in Sect. 6.2.

6.1.2 Variation of Pore Dimensions

The laser power P (measured after the focussing optics as an average over 1 s), used to produce pores in both inked and uninked skin, was varied to determine its effect on the dimensions (diameter and depth) of the pores. The diameter of pores produced in both lens and fibre set-ups increased with increasing laser power (Fig. 6.3) and, in general, the pores became deeper. Using the lens set up, the dimensions of pores formed in inked and uninked skin were similar at the same power.

A threshold effect was apparent with respect to laser power, with different poration thresholds observed for pores produced using the lens and fibre set-ups. The threshold was the highest, at 0.7 W, for the production of pores in uninked skin. Below this threshold, no effect of laser illumination, under the same exposure conditions, was observed. The use of ink on the surface of the skin brought this threshold down to 0.2 W for pores produced using the lens set-up.

The use of the fibre set-up facilitated a further reduction in the threshold for poration to 0.05 W. The maximum power achievable at the surface of the skin was lower using the fibre set-up (0.4 W) than that using the lens set-up due to coupling losses when aligning the laser beam with the fibre. Pores were therefore not observed in uninked skin using the fibre set-up because the maximum power was too low to cause poration. The pores produced using the fibre set-up were smaller in diameter than those produced using the currently available P.L.E.A.S.E.® device [2], implying that they would be quicker to heal.

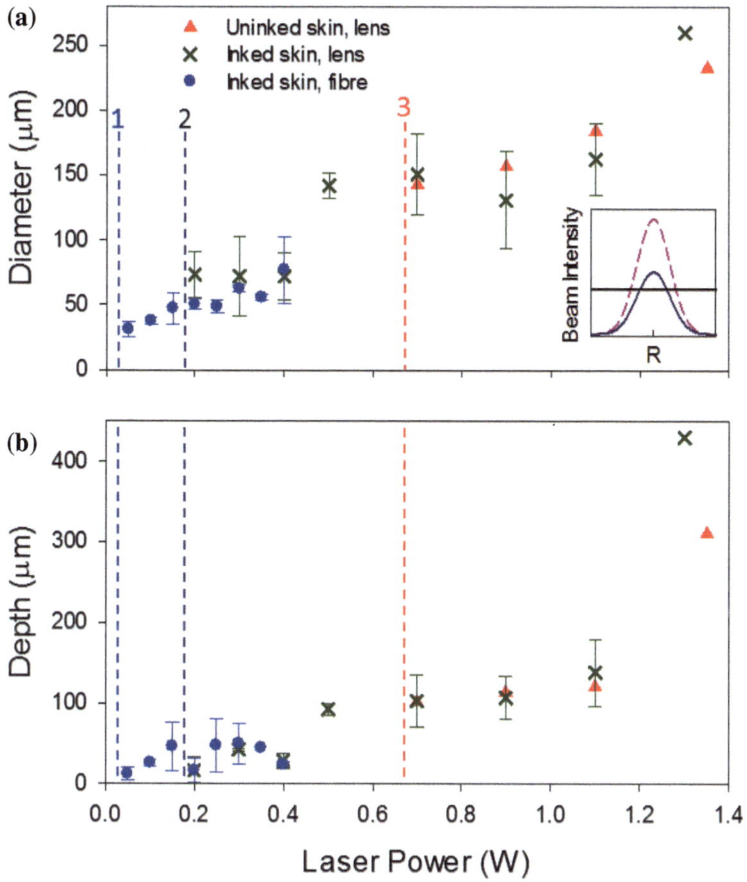

Fig. 6.3 Laser pore diameter (**a**) and depth (**b**) as a function of the laser power used. Dimensions and poration thresholds are shown for pores produced on uninked and inked skin using a lens to focus the laser light onto the skin (*red* and *green symbols* and *dashed lines 3* and *2*, respectively) and on inked skin using a fibre to direct the laser light (*blue circles* and *dashed line 1*). Exposure times of 1 s were used. Averages (±standard deviation) were calculated from three pores on inked skin, except that produced at 1.3 W, when one measurement was carried out. One pore at each power was produced in uninked skin. The distance between the skin and the lens/fibre was kept constant. Beam intensities at high (*dashed pink*) and low powers (*blue*) at a given axial distance, calculated using Eq. 6.3, as a function of radial distance from the centre of the beam, are shown in the inset. A threshold value is shown (*black line*) to compare the diameter of pores at high and low powers

6.1.3 Beam Propagation

Laser microporation using the lens set-up was not easy to reproduce. The incidence of poration was unreliable; the equipment would appear to be set up in the same way as previously but sometimes no poration was observed. This is likely to be due to the difficulty of accurately focussing the beam on the rough surface of the skin.

Laser beams often occur in the form of Gaussian beams [12]. Laser light focussed by a lens converges to its lowest diameter at the beam waist. It then diverges, at an angle which is dependent on the numerical aperture (NA) of the lens (in this case, 0.4), as the axial distance from the lens increases. A beam waist with a small radius can be obtained by focussing a laser beam using a lens with a high NA. The Gaussian beam emitted by the fibre initially has a beam radius dependent on the fibre's core diameter and then diverges according to the fibre's NA (0.04). The beam propagation from the lens and from the fibre has been modelled using the following equations.

The fluence of laser poration, the energy per unit area per exposure, is calculated using the following relation,

$$fluence = \frac{laser\ pulse\ energy}{effective\ focal\ spot\ area} \tag{6.1}$$

Fluence is commonly reported in similar experiments as it provides a measure of the safety of the exposure [12]. Due to the challenge of locating the exact position of the skin in relation to the beam waist in the case of the lens (and the large divergence of the beam away from the waist), the effective focal spot area is difficult to determine with high accuracy. For the results shown in Fig. 6.3, the distance between the lens and the outermost surface of the skin was kept constant, at approximately 2 mm, while the power was varied. The variation in power is therefore proportional to the variation in fluence.

The radius ω at which the intensity of a Gaussian beam drops to $\frac{1}{e^2}$ of its axial value as a function of the axial distance z is given by Eq. 6.2 [12],

$$\omega(z) = \omega_0\sqrt{1 + \left(\frac{z}{z_R}\right)^2} \tag{6.2}$$

where the minimum radius, the waist size, $\omega_0 = \frac{\lambda}{nNA}$, where NA is the numerical aperture of the lens, λ is the wavelength of the light and the Rayleigh range $z_R = \frac{\pi\omega_o^2}{\lambda}$. The beam emitted from the end of the fibre is assumed to have a waist size equal to the radius of the core of the fibre, 8 μm. The intensity I of a Gaussian beam at an axial distance z and radial distance r is related to the intensity at the centre of the beam at its waist I_0 [12]:

$$I(r, z) = I_0\left(\frac{\omega_0}{\omega(z)}\right)^2 \exp\left(\frac{-2r^2}{\omega^2(z)}\right) \tag{6.3}$$

where

$$I_0 = \frac{P_{peak}}{\pi\omega_0^2} \tag{6.4}$$

and the peak power for soliton pulses, as used in these experiments, $P_{peak} = 0.88\frac{P}{R\tau}$, where R is repetition rate and τ is pulse duration [12].

Beam propagation from the lens and from the fibre, calculated using Eqs. 6.2, 6.3 and 6.4, is shown in Fig. 6.4a. Due to the existence of a threshold for ablation and the divergence of the beams (Fig. 6.4a), there is a limited range of distances over which poration can occur. Before or after this range, the fluence is below the threshold for ablation. Poration was therefore heavily dependent on the position of the skin relative to the lens but less so when the fibre was used, due to its lower NA. Finding the focus of the lens (and the small range of distance for ablation) was difficult due to the skin's low absorption and significant scattering at 532 nm. Positioning the focus optimally on the skin for ablation was also complicated by the skin's roughness, caused by larger furrows in the range of 20–100 μm, and shorter and shallower secondary furrows (5–40 μm), as well as smaller scale structures [13].

The fibre set-up ensured the reliable incidence of microporation and more reproducible results than the lens set-up. Light diverges with a lower angle from the end of the fibre than the light beam diverging from the beam waist of the lens set-up. The intensities of the laser beam at distances of 30 and 100 μm from the beam waist and from the end of the fibre are shown in Fig. 6.4b. There is a slower decrease in beam intensity with distance from the fibre, allowing a greater range of distances over which poration can occur. The roughness of the skin and its position therefore influence poration less using the fibre set-up than the lens set-up.

The diameter of the modelled beam at these distances is less than the diameter of the pores observed in these experiments (Fig. 6.3). The strong scattering of light by the skin [10] is likely to have increased the effective diameter of the beam. Optical scattering of light by the skin arises from spatial variation of the refractive index within the skin due to its inhomogeneous structure. The distribution and size of the components of the skin (lipids, proteins, etc.) provide a highly scattering medium for light with a wavelength of 532 nm [14]. The situation is not easy to model due to the contribution of various factors. Each pulse is thought to ablate a certain amount of material, implying that subsequent pulses will ablate material with a different spatial structure to the previous one (i.e., as the pore depth increases). The increase in temperature associated with the formation of a plasma (as discussed in Sects. 2.4.2 and 6.4) also affects the optical absorption of skin [14]. The third factor complicating the interaction of the laser beam with the skin is the deviation of the edges of the beam from a Gaussian function at high fluences [10]. Modelling the beam propagating from the fibre and from the lens, without incorporation of the scattering coefficient, provides a measure of the beam properties as functions of the distance from the beam waist to the skin and of the laser power. Although the pore diameters were greater than the calculated beam waist diameters, the variation in pore diameter with laser power and fibre-laser distance (further described in Sect. 6.4) were successfully accounted for.

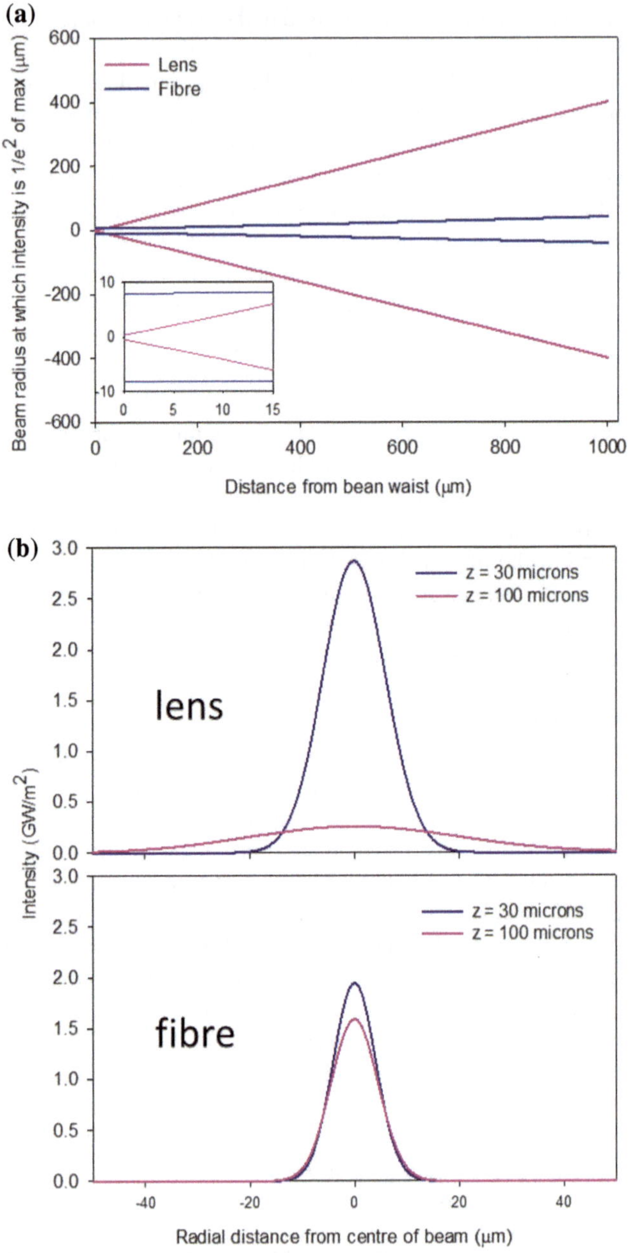

Fig. 6.4 Gaussian beam divergence using the lens and fibre set-ups, calculated using Eq. 6.3: **a** beam radius as a function of axial distance from the lens (NA = 0.4) waist and from the end of the fibre (NA = 0.04), **b** intensity as a function of radial distance at axial distances of 30 and 100 μm using the lens and fibre set-ups. Intensity was calculated using the maximum power (1.3 W for lens, 0.4 W for fibre) and the cross-sectional areas of the beam waist of the lens, and of the fibre core, respectively

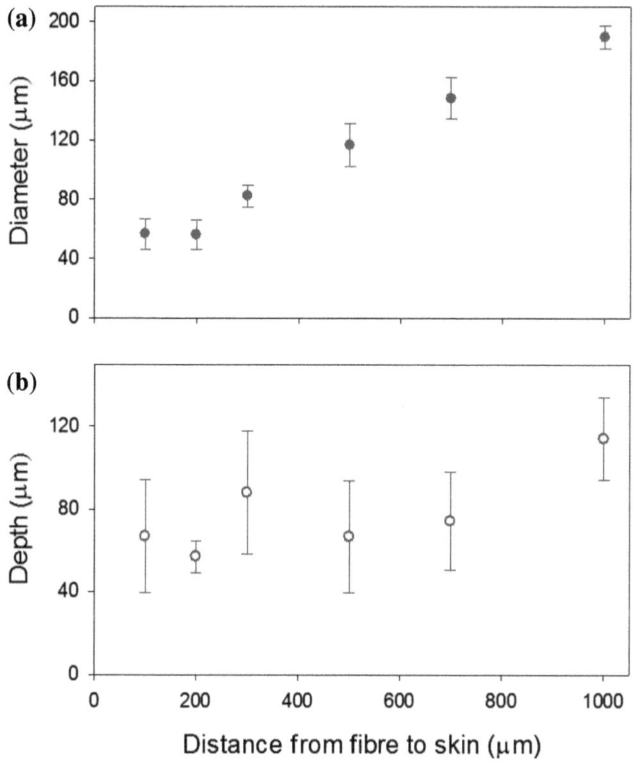

Fig. 6.5 Laser pore diameter (**a**) and depth (**b**) as a function of the distance between the surface of the skin and the fibre used to direct laser light onto the skin. Exposure times of 1 s were used and power was kept constant at 0.3 W. Averages (±standard deviation) were calculated using five pores at each power on two porcine skin samples

6.1.4 Poration Using the Fibre Set-Up

A low power for laser microporation, achievable using the fibre set-up, is useful in lowering the cost of the process. At lower powers, less heat is produced in the laser system so fewer components are needed to cool the system. Lower power emitted from the laser system would also improve its safety profile.

Once the fibre set-up was established, enabling reliable and predictable microporation to be achieved, further investigation of the pore dimensions were undertaken. The distance between the end of the fibre and the surface of the skin, and the amount of time that the skin was exposed to the laser beam, were varied.

Increasing the distance between the end of the fibre and the surface of the skin beyond 200 μm resulted in pores with larger diameters (Fig. 6.5). The surface of the skin was considered to be the outermost edge of tissue, observed using an optical microscope. As mentioned previously, the beam emitted from the fibre diverges slowly, due to its low NA. As the distance between the fibre and the skin increases, the diameter of the diverging beam becomes larger, resulting in pores with greater diameters. Even though the fluence is smaller at a greater distance, poration is still possible at 1 mm away from the skin. There may be a threshold effect at greater distances when the fluence becomes too low for poration of inked skin.

There was no significant difference between the diameter of the pores at distances of 100 and 200 μm. There is little difference between the diameters of the beam at these distances too (Fig. 6.4a). Below 200 μm, pores of the same diameter are produced regardless of the roughness of the skin, which is prominent at these small distances. The diameter of the pores is therefore predictable regardless of the roughness of the skin, which should be useful when designing a device control drug delivery through this route.

The depth of the pores did not vary significantly with distance from the fibre to the skin although the error on these measurements is quite large. The variability may be due to inherent differences in the depths of the pores or to the method of measurement. Large variations in pore depth may occur because of the interaction of the laser light with hairs in the skin, for example, which absorb laser energy in a different way due to their pigmentation and structural characteristics. There may also be other regional variations in skin structure which influence the depth of the ablated pore. However, it is likely this would also affect the diameter of the pores although this appears not to be the case.

To measure depth, the top of a pore was brought into focus using an optical microscope. The vertical position of the skin was then adjusted until the bottom of the pore was in focus. The depth of the pore was determined as the difference in vertical position between the two foci. When the pores were large (for example, at large fibre-skin distances), the top of the pore was sometimes higher on one side than the other due to the roughness of the skin at these larger scales. It was therefore difficult to determine from which point the depth should be measured. The uppermost top edge was usually used, which may have led to an overestimation of the pore depth. Previous studies have used histology to determine the depth of laser pores and this presents an alternative [15, 16], and potentially more accurate, method. This does, however, require treatment and staining of the skin, which may affect the properties of the tissue and therefore the results observed.

There was no significant difference in the diameter or depth of the pores with exposure times in the range of 0.07–1 s (Fig. 6.6). This is in disagreement with previous studies [1, 17], which report that pore depth increased with the number of pulses used for poration. These studies, however, used much lower numbers of pulses than were obtainable in these experiments; 1–50, as opposed to the 1.4–20 million pulses achieved here. The pulse durations used in these previous studies were also longer, in the range of microseconds, suggesting that a different mechanism of poration may be operative. It is possible that a relation exists, at lower exposure times, between the dimensions of the pores and the exposure time used to produce them. The poration process is likely to have been completed more quickly than the shortest exposure time used, 0.07 s. The same depths were therefore observed for longer exposure times.

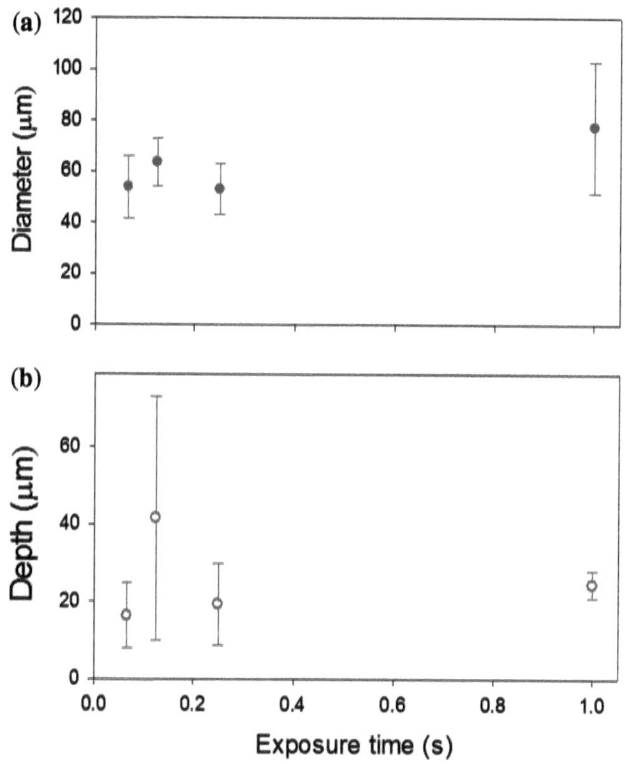

Fig. 6.6 Laser pore diameter (**a**) and depth (**b**) as a function of the exposure time. Results are shown for pores produced on inked skin using a fibre to direct the laser light. The distance between the skin and the fibre, and the laser power (0.3 W), were kept constant

6.2 Raman Micro-Spectroscopy

Raman spectroscopy proved a useful tool to objectively determine thermal damage induced during poration via the fluorescence emitted by burnt tissue.

6.2.1 Thermal Damage Detection

Raman spectra of porcine skin, which had suffered increasingly severe thermal damage, were acquired (Fig. 6.7). The skin was burnt using a soldering iron at different temperatures. The spectrum of skin burnt at 420 °C showed a large background fluorescence which was 5 orders of magnitude greater than the signal from unburnt skin. Some spectral features, characteristic of unburnt skin, were observed in skin which had been burnt at a lower temperature (160 °C), where the background signal

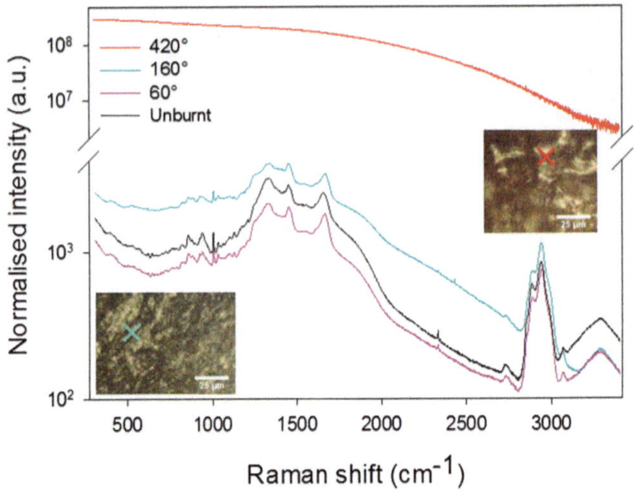

Fig. 6.7 Raman spectra of porcine skin heated to 60, 160 and 420 °C and skin which has not been burnt. Spectra have been normalised according to the laser power used for acquisition. An exposure time of 10 s to the Raman excitation laser beam was used for acquisition of all spectra. Insets show the appearance of the skin after burning at 160 and 420 °C with *cyan* and *red crosses* respectively identifying the area from which the spectrum was acquired. Scale bars are 25 μm

was still higher than that of unburnt skin. The fluorescence observed in burnt skin is thought to arise from the denaturation of collagen or other cellular proteins during heating [18].

The ability of Raman spectroscopy to identify thermal damage is again shown in the variation of spectra with increasing distance from a burnt area (Fig. 6.8). Fluorescence decreases and characteristic skin spectral features become increasingly apparent further from the burnt site. Raman spectroscopy is therefore an effective tool to determine the extent of the damage to tissue after laser poration.

6.2.2 *Comparison of Pores in Inked and Uninked Skin*

Raman spectra from the bottom of pores produced using the lens set-up in both uninked and inked skin, with depths of 300 and 120 μm, respectively, were acquired (Fig. 6.9). In uninked skin, the spectrum from the bottom of the pore shows a high background fluorescence and lack of characteristic spectral features. The spectrum is similar to that of skin burnt at 420 °C (Fig. 6.7), suggesting that the pore has been burnt, consistent with optical microscope images (Fig. 6.1). The spectrum of the skin at the bottom of the pore in inked skin is approximately 50 times lower in intensity than that in uninked skin, suggesting low thermal damage. Characteristic spectral

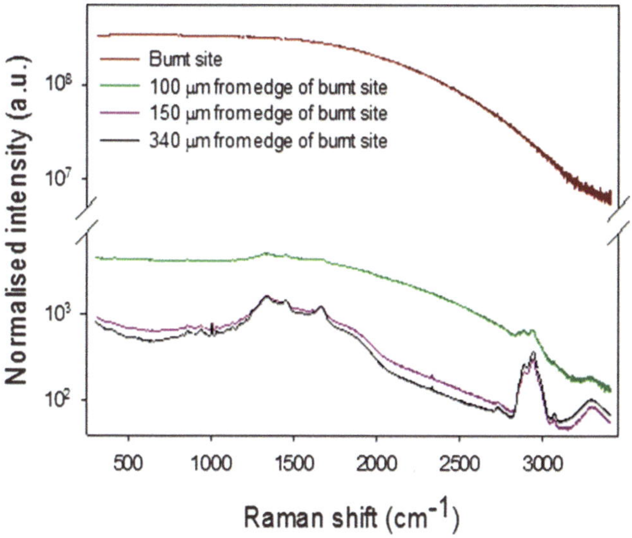

Fig. 6.8 Raman spectra as a function of distance from an area of porcine skin heated to 420 °C. Spectra have been normalised according to the laser power used for acquisition. Exposure times of 12 s to the Raman excitation laser beam were used for all spectra

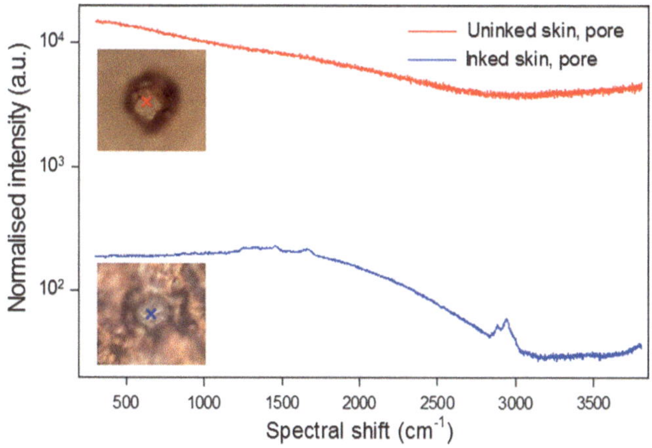

Fig. 6.9 Raman spectra of the *bottom* of laser pores produced in uninked (*red line*) and inked (*blue line*) skin. Powers of 1.3 and 0.7 W were used for poration, respectively. Spectra have been normalised according to the laser power used for their acquisition (0.5 and 100 % of the maximum power available for uninked and inked skin, respectively) using the Raman microscope. An exposure time of 120 s was used for acquisition of Raman spectra. *Insets*: optical images with *red* and *blue crosses* show the positions from which spectra were taken in uninked and inked skin, respectively

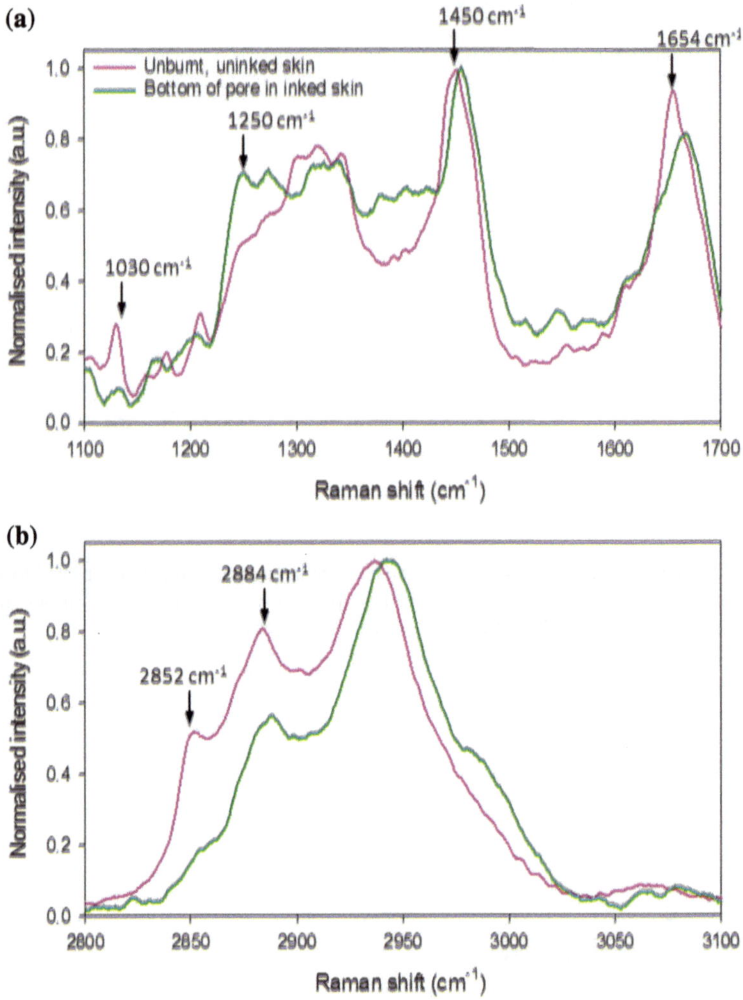

Fig. 6.10 Raman spectra of unburnt, uninked skin (*pink*) and of the *bottom* of a pore in inked skin (*green*). Background fluorescence has been subtracted to better compare the spectra. The spectrum from the bottom of the pore has been smoothed. Spectra have been normalised according to the maximum intensity within the ranges shown: **a** 1100–1700 cm^{-1} and **b** 2800–3100 cm^{-1}. *Arrows* indicate spectral features of interest

peaks are also visible. This effect was observed for spectra of three pores in both inked and uninked skin.

The fluorescent background was subtracted from the spectrum of the bottom of the pore in inked skin and the resulting difference was compared to that of uninked, undamaged skin (Fig. 6.10). The following observations were made: a decrease in the intensity of peaks related to the conformational order of the lipids within the skin (peaks at 1130, 2852 and 2884 cm^{-1}), shifts of peaks associated with the scissoring

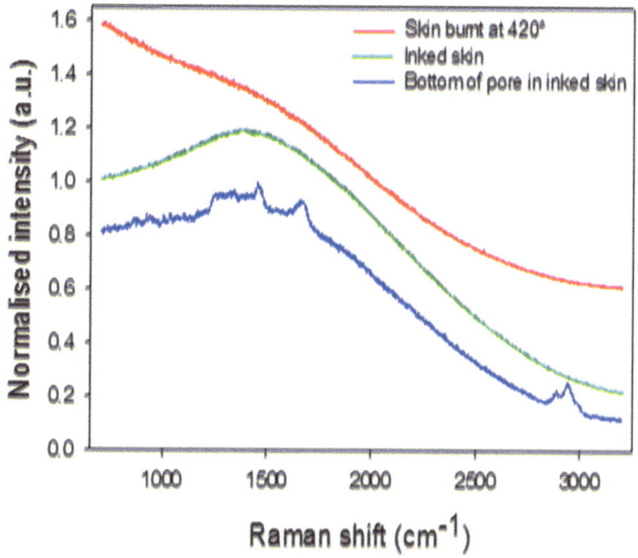

Fig. 6.11 Raman spectra of porcine skin which was heated to 420 °C (*red line*), which was inked and not heated (*green*) and which was inked then porated (*blue*). Spectra have been normalised according to maximum intensity and offset for comparison

of CH_2 and C = 0 bond stretching (1450–1456 and 1654–1670 cm^{-1}, respectively), and the appearance of a new peak at 1250 cm^{-1} assigned to the CH_2 bond wagging [19]. Previously recorded spectra of undamaged skin from locations beneath the SC show similar features [19]. It follows that, in this case, laser poration of the skin has caused little thermal damage.

The fluorescence observed in the Raman spectrum of the bottom of the pore in inked skin may, in fact, be due to the ink. Spectra of unburnt, inked skin and skin burnt at 420 °C (Fig. 6.11) show differences in the fluorescence from these two samples. The spectral shape of the fluorescence arising from ink agrees more closely with the spectrum from the pore in inked skin than the spectrum from burnt skin. This implies that the fluorescence observed in the pore may arise from residual ink, as opposed to thermal damage.

The difference between the Raman spectra of pores produced in uninked and inked skin may be due to the lower power used to produce the pore in inked skin and/or the effect of the ink on the poration mechanism. The presence of the ink permits poration to be achieved at lower power than in uninked skin. Regardless of whether the extent of thermal damage arises from the presence of the ink or from the power used, there is less thermal damage in the pore produced in inked skin. Using the fibre set-up allows even lower powers to be used to porate the skin, which would result in further reduced thermal damage.

6.3 Permeation Enhancement

To determine whether the pores created using femtosecond laser pulses would facilitate delivery in inked skin, the permeation of caffeine was investigated. Arrays of 18 × 18 pores in inked skin were produced at 0.06 and 0.3 W (Fig. 6.12) (low and high power, respectively). The diameters of the pores were 39 ± 5 and 52 ± 12 μm, and the distances between the pores were 130 ± 8 and 135 ± 18 μm for low and high power, respectively. The depths of these pores were not measured as this would have caused too great a delay between pore creation and the start of the permeation experiments. However, based on previous experiments (Fig. 6.3), it is likely that the depths of pores produced at high and low laser powers were approximately 50 and 15 μm, respectively.

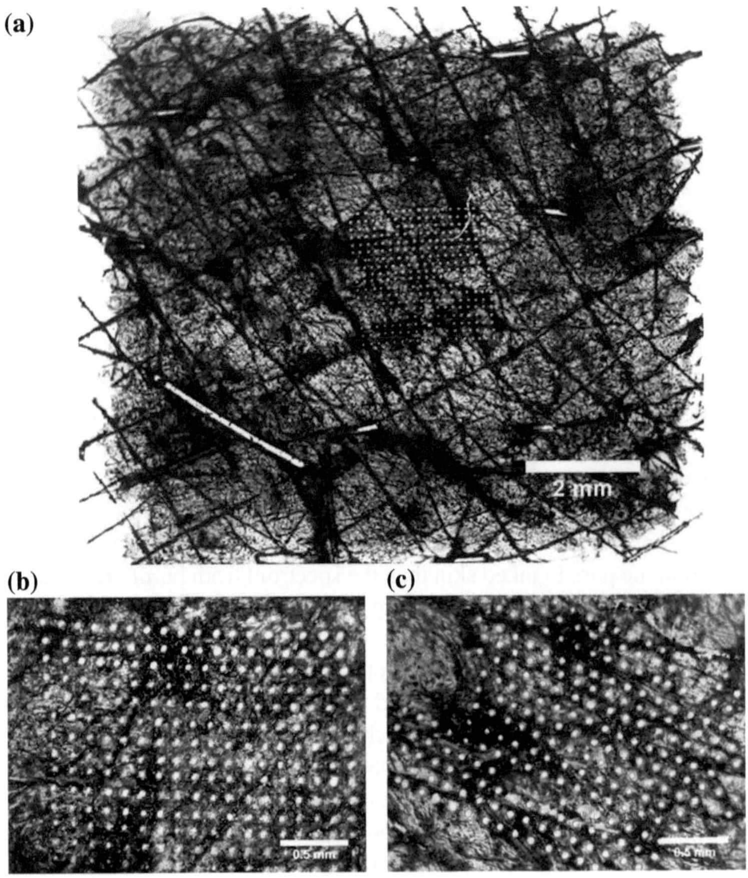

Fig. 6.12 Optical images of laser pores produced in inked porcine skin: **a** an 18 × 18 array of pores produced at a laser power of 0.06 W, **b** higher magnification of the previous image, **c** pores produced at a laser power of 0.30 W. Scale bars for **b** and **c** represent 0.5 mm

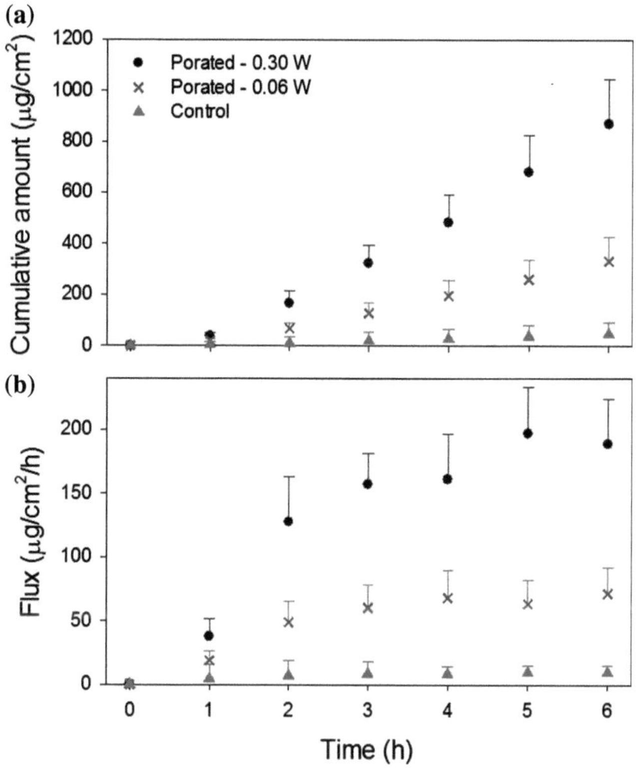

Fig. 6.13 Permeation profiles of caffeine across intact porcine skin (n = 6) and skin porated laser powers of 0.06 (n = 3) and 0.30 W (n = 3): **a** cumulative amount permeated, and **b** flux as a function of time

The cumulative passive permeation of caffeine across non-porated skin (Fig. 6.13) after 6 h was $48 \pm 44\,\mu g\ cm^{-2}$. Laser poration elicited a significant enhancement, increasing the cumulative amounts of caffeine delivered to 330 ± 100 and $870 \pm 180\ \mu g\ cm^{-2}$ for low power and high power, i.e., an enhancement factor of 7 for low power and 18 for high. The flux of caffeine through the porated skin after 6 h is 7.5 and 19.9 times greater than that through non-porated skin for low and high power, respectively. The corresponding flux increases quickly with time then begins to plateau at approximately 50 and 200 $\mu g\ cm^{-2}\ h^{-1}$.

The total area of poration produced at higher power was approximately 1.8 times greater than that at lower power. The difference in caffeine permeation between high and low power is likely due to the larger areas of the pores, as well their greater depth, when using higher power. The difference is not exactly proportional to the areas of the pores, probably because complete disruption of the SC was not achieved in all pores. As permeation increases with higher laser power, it is likely that little thermal damage, and therefore very little coagulation, has occurred, in contrast to earlier observations using older laser models [20].

6.4 Mechanism of Ablation

Due to the ultrashort (femtosecond) pulse durations used in these experiments, it is likely that the mechanism of skin poration is plasma-mediated ablation [21], in which the plasma is formed by laser induced breakdown of the tissue. Within the plasma volume, tissue vaporization and disintegration occurs [14]. For breakdown and plasma formation to occur in tissue, a threshold free electron density must be reached [22] (see Sect. 2.4.2). The presence of a threshold of poration, as shown in Fig. 6.3, is a result of this minimum electron density requirement for laser induced breakdown of the skin; the parameters of the laser irradiation must be such that the electron density builds up despite the loss of free electrons through diffusion and recombination, and loss of energy through collisions with ions [14]. This critical density can be produced by avalanche ionization and multiphoton ionization [14].

It has been shown that the optical breakdown in water is similar to that in transparent biological media [14]. The rate of pure multiphoton ionization of water is independent of the number of free "seed" electrons, as each atom is independently ionized [14, 22] and multiphoton ionization is independent of the linear absorption coefficient [23]. As the thresholds found in these experiments vary according to the presence of ink, it is likely, therefore, that laser induced breakdown occurs via avalanche ionization, as opposed to multiphoton. The laser radiation is absorbed by the black ink, resulting in thermionic emission of free electrons [14]. Once the irradiance is high enough to produce a seed electron, the necessary density of electrons for laser induced breakdown is rapidly reached. This effect has previously been studied in dye-enhanced pulsed laser ablation of enamel [23].

In uninked skin, seed electrons for breakdown may arise from impurities within the skin or chromophores which are absorbant at the laser wavelength (532 nm in this case) such as melanin. There may also be multiphoton initiated avalanche ionization. This also occurs in the presence of ink, but in that case, ionization due to the seed electrons would dominate. The absorption of laser radiation by uninked skin is much lower. A higher fluence is therefore required to ensure that a threshold electron density is achieved. The greater fluence required results in more collateral thermal damage in the surrounding tissue [24], and is consistent with the greater thermal damage observed using Raman spectroscopy (Fig. 6.9).

By increasing the laser power for poration, a greater area of the beam (at a given distance from the lens focal point or the end of the fibre) has a fluence large enough for ablation (inset of Fig. 6.3) and therefore larger pores are produced. This also explains the increasing diameters of the pores with greater distances between the fibre and the skin (Fig. 6.5); the width of the Gaussian beam increases with distance from the end of the fibre. For the data shown in Fig. 6.5, the fluence was great enough to cause poration at all distances investigated and the diameter increased with increasing distance. It is likely that, beyond a certain distance, a greater proportion of the beam will be below the threshold fluence and the diameter of the pores would then decrease with increasing distance.

Pores were shallower when produced with lower laser powers/fluences (Fig. 6.3) but the variation in depth is less pronounced at lower powers. At higher fibre-skin distances, it would be expected that the depth of the pores created would be shallower due to the decrease in fluence. Although this relation was not observed here at the powers used with the fibre set-up (Fig. 6.5), the absence of an observable trend may be due to the measurement difficulties, discussed in Sect. 6.1.4, or by the diminishing influence of the ink as the pores become deeper. The explosive ejection of the plasma per pulse would leave some material at the edges and at the bottom of the produced pore. Some of this material is likely to contain some of the absorbent ink. Its effect on the initiation of a plasma would therefore translate to lower depths with decreasing influence. Once the influence of the ink is removed, ablation is no longer possible and poration ceases.

Pores in inked and uninked skin have the same diameter and depth when produced at the same laser power above 0.7 W (Fig. 6.3). Under these conditions, a different mechanism of poration may dominate where the bulk material properties have a greater influence on poration than the surface ink. For example, the ink may be rapidly removed at these powers and the poration is achieved by the interaction of the laser radiation with uninked skin.

In the plasma mediated ablation of enamel, it was found that the ablation depth increased with each pulse [24]. In our work on inked skin, ablation depth and diameter show no relation with increasing pulse number (exposure time) (Fig. 6.6). This may be because the poration process stops at times less than the minimum used (0.07 s). Once the influence of the ink is removed, uninked skin remains that does not porate at the fluences used on inked skin. The poration process therefore stops.

6.5 Summary and Conclusions

Femtosecond pulsed laser radiation has been used to porate porcine skin and to enhance drug permeation. The application of ink to the surface of the skin lowered the power required to porate the skin. The lower powers resulted in less thermal damage, as determined using Raman spectroscopy, to the tissue surrounding the pores. This was beneficial for enhancing drug permeation as little coagulation occurred to provide a barrier to penetration.

The laser beam was focussed on or directed at the skin using either a lens or a fibre. The threshold for poration was lowest in inked skin when the fibre set-up was used. The effect of laser power, exposure time and distance between the end of the fibre and the skin on pore dimensions was attributed to the beam propagation and plasma-mediated mechanism of ablation.

This method of poration presents the following advantages over the more frequently used ablation using longer pulses, which rely on photothermal ablation and the strong absorption of the laser wavelength by water. The short pulse durations used in these experiments and the incidence of poration at low fluences ensured that there was little damage to the surrounding tissue. The use of a fibre laser and a fibre to direct the light towards the skin also increases the flexibility of the system and renders a putative poration device more compact and portable. While photothermal ablation is more efficient for strongly absorbing tissues [23], femtosecond ablation is independent of the natural variations in the tissue properties, a particular advantage given the structural heterogeneity of skin. However, femtosecond pulsed lasers are expensive. Once the technology has developed and components become more readily available, femtosecond ablation could provide an advantageous alternative to that currently used. This kind of ablation may also prove promising in applications such as damage-free separation of tissue, for example, for skin grafts [10], and for drug delivery into the nail [25].

References

1. R. Weiss, M. Hessenberger, S. Kitzmuller, D. Bach, E.E. Weinberger, W.D. Krautgartner, C. Hauser-Kronberger, B. Malissen, C. Boehler, Y.N. Kalia, J. Thalhamer, S. Scheiblhofer, Transcutaneous vaccination via laser microporation. J. Controlled Release **162**(2), 391–399 (2012)
2. Y.G. Bachhav, A. Heinrich, Y.N. Kalia, Controlled intra- and transdermal protein delivery using a minimally invasive erbium:yag fractional laser ablation technology. Eur. J. Pharm. Biopharm. **84**(2), 355–364 (2013)
3. X.Y. Chen, D. Shah, G. Kositratna, D. Manstein, R.R. Anderson, M.X. Wu, Facilitation of transcutaneous drug delivery and vaccine immunization by a safe laser technology. J. Controlled Release **159**(1), 43–51 (2012)
4. W.R. Lee, S.C. Shen, M.H. Pai, H.H. Yang, C.Y. Yuan, J.Y. Fang, Fractional laser as a tool to enhance the skin permeation of 5-aminolevulinic acid with minimal skin disruption: A comparison with conventional erbium:yag laser. J. Controlled Release **145**(2), 124–133 (2010)
5. E.M. Graber, E.L. Tanzi, T.S. Alster, Side effects and complications of fractional laser photothermolysis: experience with 961 treatments. Dermatol. Surg. **34**(3), 301–307 (2008)
6. C. Gomez, A. Costela, I. Garcia-Moreno, F. Llanes, J.M. Teijon, D. Blanco, Laser treatments on skin enhancing and controlling transdermal delivery of 5-fluorouracil. Lasers Surg. Med. **40**(1), 6–12 (2008)
7. C. Gomez, A. Costela, I. Garcia-Moreno, F. Llanes, J.M. Teijon, M.D. Blanco, Skin laser treatments enhancing transdermal delivery of ala. J. Pharm. Sci. **100**(1), 223–231 (2011)
8. G. Nicolodelli, D.P. Angarita, N.M. Inada, L.F. Tirapelli, V.S. Bagnato, Effect of photodynamic therapy on the skin using the ultrashort laser ablation. J. Biophotonics (2013)
9. K.S. Frederickson, W.E. White, R.G. Wheeland, D.R. Slaughter, Precise ablation of skin with reduced collateral damage using the femtosecond-pulsed, terawatt titanium-sapphire laser. Arch. Dermatol. **129**(8), 989–993 (1993)
10. H. Huang, Z.X. Guo, Human dermis separation via ultra-short pulsed laser plasma-mediated ablation. J. Phys. D-Appl. Phys. **42**(16) (2009)
11. G. Humbert, J.C. Knight, G. Bouwmans, P.S. Russell, D.P. Williams, P.J. Roberts, B.J. Mangan, Hollow core photonic crystal fibers for beam delivery. Opt. Express **12**(8), 1477–1484 (2004)
12. R. Paschotta, *Encyclopedia of laser physics and technology* (Wiley-VCH, Weinheim, 2008)
13. L. Tchvialeva, H. Zeng, I. Markhvida, D.I. McLean, H. Lui, T.K. Lee, Skin Roughness Assessment. InTech (2010)
14. A. Vogel, V. Venugopalan, Mechanisms of pulsed laser ablation of biological tissues. Chem. Rev. **103**(2), 577–644 (2003)
15. Y.G. Bachhav, S. Summer, A. Heinrich, T. Bragagna, C. Bohler, Y.N. Kalia, Effect of controlled laser microporation on drug transport kinetics into and across the skin. J. Controlled Release **146**(1), 31–36 (2010)

16. J. Yu, Y.G. Bachhav, S. Summer, A. Heinrich, T. Bragagna, C. Bohler, Y.N. Kalia, Using controlled laser-microporation to increase transdermal delivery of prednisone. J. Controlled Release **148**(1), E71–E73 (2010)
17. E.H. Taudorf, C.S. Haak, A.M. Erlendsson, P.A. Philipsen, R.R. Anderson, U. Paasch, M. Haedersdal, Fractional ablative erbium yag laser: histological characterization of relationships between laser settings and micropore dimensions. Lasers Surg. Med. **46**(4), 281–289 (2014)
18. N. Iftimia, R.D. Ferguson, M. Mujat, A.H. Patel, E.Z. Zhang, W. Fox, M. Rajadhyaksha, Combined reflectance confocal microscopy/optical coherence tomography imaging for skin burn assessment. Biomed. Opt. Express **4**(5), 680–695 (2013)
19. C.H. Xiao, C.R. Flach, C. Marcott, R. Mendelsohn, Uncertainties in depth determination and comparison of multivariate with univariate analysis in confocal raman studies of a laminated polymer and skin. Appl. Spectrosc. **58**(4), 382–389 (2004)
20. S.L. Jacques, D.J. Mcauliffe, I.H. Blank, J.A. Parrish, Controlled removal of human stratum-corneum by pulsed laser. J. Invest. Dermatol. **88**(1), 88–93 (1987)
21. X.H. Hu, Q.Y. Fang, M.J. Cariveau, X.N. Pan, G.W. Kalmus, Mechanism study of porcine skin ablation by nanosecond laser pulses at 1064, 532, 266, and 213 nm. IEEE J. Quantum Electron. **37**(3), 322–328 (2001)
22. D.X. Hammer, R.J. Thomas, G.D. Noojin, B.A. Rockwell, P.K. Kennedy, W.P. Roach, Experimental investigation of ultrashort pulse laser-induced breakdown thresholds in aqueous media. IEEE J. Quantum Electron. **32**(4), 670–678 (1996)
23. A.A. Oraevsky, L.B. DaSilva, A.M. Rubenchik, M.D. Feit, M.E. Glinsky, M.D. Perry, B.M. Mammini, W. Small, B.C. Stuart, Plasma mediated ablation of biological tissues with nanosecond-to-femtosecond laser pulses: relative role of linear and nonlinear absorption. IEEE J. Sel. Top. Quantum Electron. **2**(4), 801–809 (1996)
24. R. Esenaliev, A. Oraevsky, S. Rastegar, C. Frederickson, M. Motamedi, Mechanism of dye-enhanced pulsed laser ablation of hard tissues: Implications for dentistry. IEEE J. Sel. Top. Quantum Electron. **2**(4), 836–846 (1996)
25. J. Neev, J.S. Nelson, M. Critelli, J.L. McCullough, E. Cheung, W.A. Carrasco, A.M. Rubenchik, L.B. DaSilva, M.D. Perry, B.C. Stuart, Ablation of human nail by pulsed lasers. Lasers Surg. Med. **21**(2), 186–192 (1997)

Chapter 7
Conclusion

The delivery of topically applied compounds can be enhanced by either optimising the formulation in which they are applied or by removing the SC, which provides a rate limiting barrier. AFM and Raman micro-spectroscopy have been used to investigate polymeric films for (trans)dermal drug delivery of varying compositions. The effect of two different additives on the properties of polymeric films were assessed using AFM and Raman micro-spectroscopy techniques. Triethyl citrate (TEC), a commonly used plasticizer, or medium-chain triglycerides (MCT) were incorporated with betamethasone-17-valerate (BMV), a drug commonly used for inflammatory skin conditions, into Eudragit and Klucel polymer films. The homogeneity/inhomogeneity of films was observed using AFM imaging and nanoindentation was used to determine elastic moduli. Raman chemical mapping revealed the distribution of the films' components. Characterisation by these methods can lead to optimisation of the constituents of the films and their proportions.

Laser microporation was used to remove the SC and to provide a means of enhancing the delivery of (at least) low molecular weight compounds. Poration was achieved using a femtosecond pulsed laser with a wavelength of 532 nm and the application of an ink to the skin's surface. Little thermal damage was observed around the pores. A fibre laser and fibre optics were used, improving the compactibility of a possible poration device.

7.1 Summary of Results

AFM and Raman micro-spectroscopy investigations of films containing TEC supported the use of this additive to increase the flexibility of the films and act as a plasticizer. AFM images of the films showed the homogeneity of films, containing 0, 20 and 40 % TEC, and the lack of crystallization of the incorporated drug, BMV.

Nanoindentation of films of varying compositions was undertaken using spherical AFM probe tips as indenters. The deformation of the samples as a function of the load applied revealed their viscous, elastic and plastic behaviour. By varying the approach

H. Garvie-Cook, *Novel (Trans)dermal Drug Delivery Strategies*,
Springer Theses, DOI 10.1007/978-3-319-28901-4_7

rate of the indentation and by including the use of a surface delay at maximum load, viscous behaviour was observed to be greater for samples containing TEC.

Elastic moduli were calculated from nanoindentation loops using two models: the elastic Hertzian model and an adapted Oyen & Cook model, which takes into account time-dependent viscous deformation during unloading. Values of elastic modulus determined using the Oyen & Cook model were consistently higher than Hertzian values, but both models detected a significant decrease in elastic modulus with increasing TEC content.

To complement the information acquired on the homogeneity of the films using the AFM, Raman chemical mapping of Eudragit and Klucel films with 20 % TEC and 1.2 % BMV was carried out on a larger scale. Both TEC and BMV were distributed evenly over the mapped area of the films. The physical state of the drug was determined using the position of its characteristic Raman peak and was found to be more soluble in Eudragit films than in Klucel films. BMV solubility in both films was effectively constant over the mapped areas.

Changes in the films' mechanical properties when cast on the surface of the skin, as opposed to on glass slides, were determined using nanoindentation of Eudragit films containing 0, 20 and 40 % TEC. Moisture from the skin was thought to have a plasticizing effect on films without plasticizer, decreasing their elastic modulus compared to films cast on glass slides. Elastic moduli of films incorporating 20 and 40 % TEC did not change significantly as the plasticizing effect of the skin's moisture was small compared to that of the incorporated TEC.

The addition of plasticizer to the films allowed their elastic moduli to be matched to that of skin. When in place on the surface of the skin, the films will therefore successfully flex with the skin without breaking contact. This prolongs their intimate contact and maintains a constant area over which drug delivery occurs. The lack of drug crystallization observed is also important in the optimisation of these films and implies that the incorporated BMV is available to be absorbed by the skin. The composition of polymeric films for drug delivery can therefore be optimised, for prolonged contact and drug bioavailability, using a combination of AFM and Raman micro-spectroscopy techniques.

Further investigation of the interaction between polymers of differing hydrophobicities and the skin would give more information on the plasticizing effect of the skin's moisture. Klucel is a more hydrophilic polymer than Eudragit and is soluble in water, so it is likely that its interaction with the skin's moisture would have more of an effect on its mechanical properties. Raman chemical mapping could also be performed of the films on the surface of the skin. The distribution of drug and plasticizer within the films could be mapped as a function of time, leading to information on the absorption of the drug (and possibly of the plasticizer) by the skin.

The use of phase imaging would complement the topographical information acquired using the AFM. Phase imaging contrast arises from the phase shift in the oscillating cantilever when damped by a region of the sample with, for example, lower viscosity or greater adhesion. Contrast may appear where there are no obvious topographical features, such as the case for Eudragit films incorporating TEC. The AFM used in these experiments did not have phase imaging capabilities. Information was acquired on the chemical homogeneity of the films using Raman micro-spectroscopy but phase imaging would provide complementary information at the nanoscale.

The same analysis techniques were applied on films incorporating medium chain triglyceride (MCT), which was originally incorporated to determine its function as a plasticizer. AFM imaging revealed two phases in both Eudragit and Klucel films incorporating 20 % MCT, in contrast to the homogeneous films resulting from the incorporation of 20 % TEC. Crystallization of the incorporated drug was not observed in either phase. Inclusions with diameters of 0.8 ± 0.2 and 1.9 ± 0.9 μm were observed in Eudragit and Klucel films, respectively. The presence of this phase separation implied that MCT does not act as a conventional plasticizer when incorporated into polymeric films. To elucidate the nature of these inclusions, the tapping mode force used to acquire height images was increased. At higher tapping mode forces, the observed inclusions deformed more and appeared deeper as the force increased. This was due to their softer nature (lower elastic modulus and viscosity) compared to their surroundings.

AFM nanoindentation of these phase separated films revealed the differing mechanical properties of the inclusions and their surroundings. The mechanical properties of the surrounding material were similar to those in films without MCT. Inclusions, however, deformed more under a given load and showed more viscous behaviour. The mechanical properties of the inclusions was dependent on the polymer film in which they were contained, suggesting that some polymer was present within the inclusions.

Elastic moduli of the inclusions, calculated using the adapted Oyen & Cook model, were approximately 3 and 4 times lower than the Eudragit and Klucel films without MCT, respectively. The range of elastic moduli varied from that of films without MCT to the modulus of skin. The inhomogeneity of the films, however, implied that some regions are more likely to lost contact with the skin when on its surface being stretched.

Raman chemical mapping was used to determine the chemical distribution of MCT and the model drug BMV. Mapping revealed that more MCT was within the inclusions than outside. BMV was distributed more evenly than MCT, but more was found outside of the inclusions. The position of the BMV Raman peak was, again, used to map the physical state of the drug within the films. BMV was equally soluble across the mapped area of the Eudragit film but was more soluble within Klucel inclusions than outside.

In vitro release was enhanced when MCT was incorporated into both Eudragit and Klucel films. The enhancement in release from Eudragit films was attributed to the viscous and softer nature of the inclusions, providing an environment in which the drug was more highly mobile. The inclusions in Klucel films increased the solubility of BMV. The increased mobility and solubility of BMV within the Klucel inclusions was therefore thought to explain the enhancement in its release.

Although the release of BMV was enhanced, films incorporating MCT showed mechanical and chemical inhomogeneity, which may adversely affect their ability to remain in intimate contact with the skin. A compromise could be reached by incorporating the conventional and effective plasticizer TEC and MCT. This could result in a release enhancement due to MCT within the films and greater flexibility due to the incorporation of TEC.

Assessing the properties of these films when on the surface of the skin may also provide useful insights into the mechanism behind the release of BMV. The size of the inclusions was dependent on the temperature at which they formed. Solvent evaporation of the films when formed on the surface of the skin is likely, therefore, to be different than when they were cast onto glass, affecting the size and distribution of the inclusions. Raman chemical maps of MCT and BMV within the films would reveal the distribution of these components and their changes over time and could help to elucidate the release mechanism.

AFM and Raman micro-spectroscopy have been used to characterize polymeric films for (trans)dermal delivery. SC removal can be used as well as, or instead of, formulation optimisation for delivery enhancement. Poration of skin, to improve the delivery of lower molecular weight compounds and permit the delivery of compounds with molecular weights over 500 g mol^{-1}, was achieved by ablating the skin using a 532 nm laser with a pulse duration of 300 fs. The laser beam was either focussed onto the skin using a lens or directed towards the skin using a photonic crystal fibre (PCF). The skin was used either without (uninked) or with (ink) the application of a dye to its surface.

Optical microscopy revealed burnt tissue at the edges and the bottom of pores produced in uninked skin. Thermal damage in pores in inked skin was less obvious but the burnt appearance of the ink made identification of black tissue more difficult. The diameter and depth of pores increased with increasing laser power but a threshold, below which poration did not occur, was observed. The threshold was highest for the poration of uninked skin and decreased when ink was applied. Poration using the fibre set up was achievable at the lowest powers.

The lens set-up resulted in unreliable poration, thought to be due to difficulties in focussing the light on the skin and the limited range of distance from the beam waist over which poration could occur. The numerical aperture (NA) of the fibre was an order of magnitude lower than that of the lens, meaning that poration was possible over a greater range of distances between the end of the fibre and the skin. Poration using the fibre was much more reliable and reproducible. Modelling the Gaussian beam propagating from the lens beam waist and from the end of the fibre accounted for the greater pore diameters observed at increasing laser powers and distances between the end of the fibre and the surface of the skin.

Raman micro-spectroscopy was used to determine the extent of the thermal damage to the skin resulting from laser poration. Purposely burning the skin allowed the difference in the Raman spectra of damaged and undamaged skin to be determined. Fluorescence strongly increased with the extent of thermal damage and the characteristic spectral features of the skin were lost. A spectrum acquired at the bottom of a pore in uninked skin showed a large fluorescent signal, suggesting that the skin had been thermally damaged by the poration process, as observed in optical microscope images. In contrast, a spectrum from the bottom of pores in inked skin revealed spectral features characteristic of the skin and lower fluorescence intensity, which implied that the skin at the bottom of pores in inked skin was less damaged.

Once a reproducible method for poration was achieved using the fibre set up, caffeine permeation experiments were carried out. The presence of pores enhanced the permeation of caffeine. A greater enhancement was observed for pores produced using a higher power, suggesting that the tissue around the pores had not coagulated due to thermal damage. Due to the short pulse duration used in these experiments, the mechanism behind the poration observed was thought to be plasma mediated ablation. The ink on the surface of the skin provided a means by which plasma formation was initiated. Plasma formation was possible in skin without ink but required higher powers, resulting in greater thermal damage.

Thus a technique has been developed to ablate skin using a femtosecond pulsed laser with a wavelength of 532 nm. The interaction between the laser and the skin was enhanced by applying an ink to the surface of the skin, bringing the power required for ablation down. Devices are currently available for laser microporation of skin which ablate the skin photothermally, using a laser wavelength (most commonly 2940 nm) which is strongly absorbed by the water within the tissue. However, this devices are associated with adverse side effects. Here, pores produced using a femtosecond pulsed laser, directed to the skin using a fibre, were smaller than those obtainable using the devices that are currently available. There is also little thermal damage surrounding the pores. These qualities suggest that the delivery of topically applied compounds and the subsequent healing of the skin will be advantageous. The use of a fibre laser and fibre optics also improves the compatibility of a possible device to porate the skin and visible light improves its safety profile.

All experiments were undertaken on non-pigmented porcine skin. As the introduction of a black ink on the surface of the skin initiated ablation, investigations are recommended on pigmented skin, with and without the application of ink. A device designed to enhance the delivery of topically applied compounds must work effectively on skin of all types. The threshold for ablation of uninked, pigmented skin may be lower than that in non-pigmented skin. The dimensions of pores are likely therefore to differ. Ideally, the application of ink on the surface of pigmented skin would result in similar poration to that on non-pigmented skin and could therefore be used as a measure to keep poration consistent from person to person.

The use of a permanent marker on the surface of the skin may not be very appealing to people requiring treatment, especially when multiple treatments are required. The laser used for poration needs to be strongly absorbed by the dye on the surface of the skin for it to initiate ablation. Either the dye, or the way that the skin is dyed, could be varied. For example, the dye could be applied in a polymeric film, as described earlier in this work, which could be removed after poration using ethanol or water.

The ability of this poration technique to facilitate the delivery of high molecular weight compounds, such as vaccines and proteins, across the skin should also be investigated and compared to other enhancement techniques, such as iontophoresis and microneedle delivery. This poration technique also has possible applications in the delivery of compounds applied to the nail surface.

7.2 Concluding Remarks

Novel topical polymeric films for (trans)dermal delivery have been assessed at the micro- and nanoscale using AFM and Raman micro-spectroscopy. The information provided by these complementary techniques can lead towards the elucidation of release mechanisms and the selection of the films' constituents. Ablating the skin, after the application of dye, using a femtosecond pulsed laser resulted in little thermal damage to the surrounding tissue. This technique, therefore, provides a means by which the diffusion of topically applied compounds is increased. This thesis has therefore described methods by which the delivery of topically applied drugs can be enhanced and optimised.

MIX
Papier aus verantwortungsvollen Quellen
Paper from responsible sources
FSC® C105338

If you have any concerns about our products,
you can contact us on
ProductSafety@springernature.com

In case Publisher is established outside the EU,
the EU authorized representative is:
Springer Nature Customer Service Center GmbH
Europaplatz 3, 69115 Heidelberg, Germany

Printed by Libri Plureos GmbH
in Hamburg, Germany